The BIG Book of Glues, Brews, and Goos

The BIG Book of Glues, Brews, and Goos

500+ Kid-Tested Recipes and Formulas for Hands-On Learning

DIANA F. MARKS

Illustrated by
Donna L. Farrell

 LIBRARIES UNLIMITED

AN IMPRINT OF ABC-CLIO, LLC
Santa Barbara, California • Denver, Colorado • Oxford, England

Batavia Public Library
Batavia, Illinois

Library of Congress Cataloging-in-Publication Data

Marks, Diana F.
    The big book of glues, brews, and goos : 500+ kid-tested recipes and formulas for hands-on learning / Diana F. Marks ; illustrated by Donna L. Farrell.
        pages cm
    Includes bibliographical references and index.
    ISBN 978-1-61069-771-2 (pbk : alk. paper) — ISBN 978-1-61069-772-9 (ebook)   1. Chemistry, Technical—Experiments.   2. Chemistry, Technical—Study and teaching (Elementary)   3. Creative activities and seat work.
I. Farrell, Donna L.   II. Title.
    TP168.M37 2015
    660—dc23   2014027063

ISBN: 978-1-61069-771-2
EISBN: 978-1-61069-772-9

19 18 17 16 15   1 2 3 4 5

This book is also available on the World Wide Web as an eBook.
Visit www.abc-clio.com for details.

Libraries Unlimited
An Imprint of ABC-CLIO, LLC

ABC-CLIO, LLC
130 Cremona Drive, P.O. Box 1911
Santa Barbara, California 93116-1911

This book is printed on acid-free paper ∞
Manufactured in the United States of America

Contents

Acknowledgments

I want to thank my husband, Peter, and our sons, Kevin and Colin, for their support and encouragement.

I thank my good friend Beth Auwarter for her advice and knowledge regarding makerspaces.

I thank Donna L. Farrell for her artistic talents and her illustrations.

I thank Patrick, Kylie, Kelsey, and Jack. They were willing experimenters in my times of need.

Introduction

I am very fortunate. In 1996, I published *Glues, Brews, and Goos: Recipes and Formulas for Almost Any Classroom Project*. The book was well received. In 2003, I published *Glues, Brews, and Goos: Recipes and Formulas for Almost Any Classroom Project: Volume 2*. That book was also popular. Now I happily have the opportunity to incorporate the two books, add more recipes, update the material, and provide more relevance to the activities. *The BIG Book of Glues, Brews, and Goos: 500+ Kid-Tested Recipes and Formulas for Hands-On Learning* is the result of all that effort.

If you work with children, you will find this resource book invaluable. Each tried and true, safe concoction uses easily obtainable ingredients, and suggestions abound as to when and why each formula can be used. Recommendations regarding how to link recipes to curriculum and projects make the final results fun, educational, and challenging.

My goals for this book are the same as the goals for the previous two books. I want children to learn and create while having fun. These last few years education has seemed to focus on standardized testing, and somehow a child's need for playful experimentation has been left behind. I hope this book inspires time for freedom of expression and making messes.

I want children to connect with the past. Here children can learn about ancient Egyptian mummification processes. They can make and taste hardtack, the staple of both sixteenth-century sailors and Civil War soldiers. They can harken back to the old ways of making batik, and they can make a primitive telephone.

I want children to connect with nature and nature's bounty. This book provides plenty of activities to grow plants, feed birds, and observe insects. They can dry flowers and make old-fashioned pomanders. They can dry fruits and make horseradish.

I want children to explore the science of art. They can find out why certain paints work best on certain surfaces. They can figure out why glues work and when papier-mâché is better than plaster of Paris. They can discover the science behind making paper and the intricacies of brewing natural dyes.

I also want children to explore the art of science. Sometimes children think science occurs in big laboratories with fancy equipment. However, they can easily make different types of volcanic eruptions, send messages to friends with invisible inks, and investigate all kinds of bubble mixtures (outside, of course). They can play with different types of non-Newtonian fluids and make goos that snap and pop.

I want children to connect with each other, and I hope children see themselves as part of a global society. They can share homemade mints on Valentine's Day and try out Irish potato candies on Saint Patrick's Day. They can create a piñata to celebrate Cinco de Mayo, and they can give Bulgarian martenitsas to best friends.

Children have no problem connecting with food. I have shared recipes from long ago, such as pickles, biscuits, and apple butter. I show them how to make foods that they usually do not make, such as peanut butter, marshmallows, and cheese. I have included favorites, such as gingerbread houses, cookies, and fudge.

This book would make a great addition to a makerspace! Place this book next to the construction paper, glitter, and scissors, and you will see creativity blossom. Although the activities in this book are low-tech, the results can be high-tech. Kids could graph the results of their homemade weather stations, blog about crystal mixtures, and produce videos of vertical and horizontal spinners.

This book provides a plethora of ideas for many occasions. Suppose you want to study the American colonial period. You can find a chapter full of ideas. However, the table of contents can give you many other ideas, such as making journey cakes, dipping candles, and brewing mustard.

I hope you and yours enjoy this book. I hope this book becomes dog eared, stained, and marked up. I hope your young charges become the better for participating in these activities, and that they become inquisitive, accomplished, and caring learners.

Tips

Make safety the top priority. Try to anticipate any possible problems and eliminate all dangers. Review safety procedures with students. Ensure that children consume only what you want them to eat or drink.

Always test a formula before using it with children.

Carefully read through the entire activity before beginning.

Make sure all materials are assembled before beginning a project.

Keep pots, spoons, and utensils used for making food separate from those used in nonfood projects.

Keep plenty of hot pads, old cotton towels, paper towels, and wet wipes around. You can never anticipate everything when children and good times get together.

Children should wear protective clothing. Smocks, old men's shirts, lab coats, and even aprons add to the fun and make parents happy. Keep some safety goggles and rubber gloves around just to be super careful.

Throw away all unused materials in the trash. Some formulas, such as the slimes group, could clog plumbing. Make sure any heated materials have cooled before discarding them.

Recipes and formulas note when a stove or heating element is required. However, electric frying pans and pots, if available, could be more useful. Temperatures can be controlled more accurately on electric appliances, and they can be used right in the classroom. Also, electric pots and frying pans can be easily transported to a sink for cleaning.

Because hot running water is not available in many classrooms, cleaning up can be difficult. If running water is not available, use plastic bags to hold ingredients instead of mixing bowls. Students like to seal the bags and mix the contents by squeezing. Though plastic bags are not the most environmentally preferred material, their use may make the difference between carrying out a project or dismissing it as too cumbersome.

Always keep a few pens around to write in this book. Keep notes on what works and what does not work with your children.

Experiment, create, learn, and enjoy!

1
Clays and Doughs

Clays and doughs can be divided into two groups: those that dry and those that do not dry. A clay or dough that dries will retain its shape and can be used to make permanent projects. A dough that does not dry may keep its shape if left undisturbed. However, it can be used only for temporary activities. Most of the recipes included here will dry. Each recipe will indicate whether the clay will dry or not.

Except for real clay, the following clays and doughs do not require a kiln. Clays and doughs can clog a sink, so all such materials should be thrown away in the trash.

■ Silly Stuff
[Makes 2 1/4 cups—enough for 2 children]

NOTE: Similar to Play Doh® silly stuff is for temporary use. It does not dry well.

Materials

1 cup all-purpose flour	food coloring (optional)
1/2 cup salt	stove or heating element
2 tablespoons vegetable oil	pot
1 cup water	mixing spoon
2 teaspoons cream of tartar	airtight storage container

Procedure

1. Mix all ingredients in pot.
2. Cook over medium heat, stirring until mixture sticks together in a ball.
3. Remove pot from heat and let dough cool.
4. Squeeze and knead. Have fun!
5. Store in airtight container.

■ Playful Plastic

NOTE: Similar to Plasticine® playful plastic is for temporary use. It does not dry well. This clay can be used for stop-motion photography.

Materials

4 ounces beeswax, grated

5 old crayons

2 tablespoons petroleum jelly

old double boiler that will never cook food again

old mixing spoon

stove or heating element

water

wax paper

Procedure

1. Pour grated beeswax into smaller pot of double boiler.
2. Pour water into larger portion of double boiler.
3. Place smaller portion of double boiler into larger portion.
4. Place double boiler onto stove and melt wax.
5. Remove paper from crayons and add crayons to melting wax.
6. When wax and crayons have melted, remove double boiler from stove. Add petroleum jelly and let the mixture cool for 30 minutes.
7. Remove mixture from double boiler and place on wax paper. Have fun!

■ Alum Dough

[Makes 2 1/2 cups—enough for 2 children]

NOTE: This recipe may be the perfect dough. It keeps without refrigeration for a couple of months, it dries overnight, and it does not have to be cooked.

Materials

3 teaspoons alum

1 1/2 cups all-purpose flour

1 cup salt

1 cup boiling water

2 teaspoons vegetable oil

mixing bowl

mixing spoon

powdered tempera paints

mixing containers

airtight storage containers

Procedure

1. Combine dry ingredients in mixing bowl.
2. Add boiling water and oil. Mix thoroughly.
3. Divide dough into several portions, place in mixing containers, and add tempera paints to achieve desired colors.
4. Model and let dry.
5. Store unused dough in airtight containers.

■ Soap Clay
[Makes 1 1/4 cups—enough for 1 child]

NOTE: A shiny product, soap clay can be applied to other surfaces to make "snow."

Materials

3/4 cup powdered laundry soap (not detergent)
1 teaspoon warm water

mixing bowl
mixing spoon
electric mixer

Procedure

1. Mix powdered laundry soap and water in bowl.

2. Beat with mixer until it feels like clay.

3. Sculpt clay. It dries to a shiny finish.

■ Soap Dough
[Makes 3 1/4 cups—enough for 3 children]

NOTE: The color is part of the dough. It has a slightly slippery feel and does not dry well.

Materials

1/2 cup salt
2 cups all-purpose flour
1 tablespoon powdered tempera paint

1 tablespoon liquid soap
1 cup water
mixing bowl
mixing spoon

Procedure

1. Combine salt, flour, and tempera paint in mixing bowl.

2. Stir in liquid soap.

3. Slowly add water to make soft dough.

4. Create!

■ Quick Clay
[Makes 2 cups—enough for 2 children]

NOTE: As the name implies, this clay hardens quickly. It can be stored in a closed container for 1 month.

Materials

1 cup baking soda
1/2 cup cornstarch
2/3 cup warm water
pot

stove or heating element
mixing spoon
food coloring (optional)
acrylic gloss and brush (optional)

Procedure

1. Stir together soda and cornstarch in pot.
2. Add warm water and stir. Heat on medium heat until it boils. It will look like mashed potatoes.
3. Remove from stove and cool.
4. Knead clay and add food coloring if desired.
5. Shape object and let it dry.
6. Finish with acrylic gloss if desired.

■ Salad-Dressing Clay
[Makes 5 cups—enough for 3 children]

NOTE: This dough is oily, so keep lots of wipes at hand. The dough is smooth and fun to shape. It does not dry well, so children should take advantage of this dough's temporary nature and just play.

Materials

1 cup salt
3 cups all-purpose flour
1 cup water
1/3 cup vegetable oil

3 tablespoons white vinegar
mixing bowl
mixing spoon
airtight storage containers

Procedure

1. Combine all ingredients in mixing bowl.
2. Knead and sculpt.
3. Store leftovers in airtight containers

■ Glue-Shampoo Dough

[Makes 2 cups—enough for 2 children]

NOTE: This dough requires no cooking, and ingredients are easy to obtain. It coils nicely to make pots, and finished pieces can air dry.

Materials

1/2 cup white glue	mixing bowl
1/3 cup shampoo	mixing spoon
1 1/2 cups all-purpose flour	paints and paintbrushes (optional)

Procedure

1. Combine all ingredients in mixing bowl.
2. Knead dough in bowl.
3. Mold dough and let objects dry for 1 or 2 days.
4. Paint objects if desired.

■ Cornstarch-Glue Dough

[Makes 3 1/4 cups—enough for 3 children]

NOTE: Easy to make, this dough is white and pliable. It does not require cooking, but it does not dry well.

Materials

1 cup cornstarch	1 cup water
1/2 cup white glue	mixing bowl
3/4 cup all-purpose flour	mixing spoon

Procedure

1. Combine cornstarch, glue, and flour in mixing bowl.
2. Gradually add water and knead.
3. Have fun! This dough should be used the same day it is made.

■ Cornmeal Dough
[Makes 2 1/2 cups—enough for 2 children]

NOTE: Grainy and gooey, this dough will dry to a hard finish. It does not require cooking.

Materials

1 cup cornmeal   mixing bowl
1 cup flour   mixing spoon
2/3 cup salt   airtight storage container
1 cup water

Procedure

1. Combine dry ingredients in mixing bowl.
2. Add enough water to make dough.
3. Form into shapes and allow projects to air dry.
4. Store any unused dough in airtight storage container.

■ Oat Dough
[Makes 3 1/2 cups—enough for 3 children]

NOTE: Oats give this dough a rough texture. It does not dry well. Use the dough the same day it is made.

Materials

2 cups boiling water   mixing bowl
1 cup oatmeal   mixing spoon
1 cup all-purpose flour   cinnamon (optional)

Procedure

1. Combine boiling water and oatmeal in mixing bowl.
2. Add enough flour to make dough.
3. Add cinnamon for a wonderful fragrance.
4. Shape and reshape dough (e.g., children could first model a mountain ridge, then reshape dough in a plateau, then reshape dough into a river delta).

■ Pearly Clay
[Makes about 1 1/2 cups—enough for 1 child]

NOTE: This clay is very translucent when soft and fairly translucent when dry. It stores well in an airtight container. Pearly clay will harden within 24 hours.

Materials

1/2 cup salt   stove or heating element
1/2 cup boiling water   pot
1/4 cup cold water   mixing bowl
1/2 cup cornstarch   mixing spoon

Procedure

1. Pour salt into boiling water.

2. In mixing bowl, combine cornstarch and cold water.

3. Add cornstarch mixture to salt solution.

4. Cook over low heat, stirring constantly.

5. When mixture is consistency of stiff cookie dough, remove from heat and cool.

6. Knead until clay is pliable.

7. Mold into desired shapes.

8. Let pearly clay shapes dry overnight, or bake them at 200°F for 1 hour.

■ Flour Clay
[Makes 6 cups—enough for 3 children]

NOTE: Flour clay requires no cooking. It is versatile in that it can be baked or allowed to air dry. Children can make thin coils of clay and intertwine them to make baskets. Children can also make "bagels" and other bread look-alikes. Flour clay projects can last for years and can be painted after the clay has dried.

Materials

4 cups all-purpose flour   mixing spoon
1 1/2 cups warm water   cookie sheet
1 cup salt   refrigerator
mixing bowl

Procedure

1. Thoroughly mix flour, water, and salt in mixing bowl and refrigerate for 30 minutes.
2. Make relatively thin baskets, decorations, or other projects.
3. Place on cookie sheet. Air dry for several days or bake at 300°F for 1 hour.
4. Refrigerate any unused clay in a plastic bag.

■ Very Easy Flour Dough
[Makes about 3 1/2 cups—enough for 3 children]

NOTE: This clay does not keep for any length of time. Finished pieces do dry.

Materials

3 cups all-purpose flour   water
3 tablespoons powdered tem-   mixing bowl
   pera paint   mixing spoon
1/2 cup vegetable oil   wax paper

Procedure

1. Combine flour, powdered tempera paint, and vegetable oil in mixing bowl.
2. Add enough water to make soft dough.
3. Create objects and let air dry on wax paper.

■ Flour and Sugar Dough
[Makes about 2 1/2 cups]

NOTE: As clay, this material does not keep for any length of time. Finished pieces do dry.

Materials

1 cup all-purpose flour
1 cup sugar
1 cup cold water
1 cup boiling water

old pot
stove or heating element
mixing spoon
wax paper

Procedure

1. Combine flour, sugar, and 1 cup cold water in old pot.
2. Pour in boiling water and cook for 5 minutes, stirring continuously.
3. Cool.
4. Create objects and let air dry on wax paper.

■ Bread Slice Clay
[Allow 1 or 2 slices of bread per child]

NOTE: Bread slice clay has few ingredients. This clay is easy to use with an entire group. Keep wipes on hand to clean messy fingers. The clay will keep for a few days.

Materials

1 slice bread
1 teaspoon white glue
1 teaspoon water

food coloring (optional)
acrylic gloss and brush (optional)

Procedure

1. Remove crust from bread. Save the crusts for the birds.
2. Pour glue and water into center of bread.
3. Knead until a ball of dough forms (approximately 5 minutes).
4. Add food coloring if desired.
5. Sculpt clay and let harden overnight.
6. Apply acrylic gloss over piece if you wish to save it for a long time.

■ Bread Dough
[Makes about 1/2 cup]

NOTE: Only small objects can be made with this recipe. This clay does dry. Save the bread crusts for treats for the birds.

Materials

2 slices day-old bread	4 drops white vinegar
2 tablespoons white glue	bowl
3 drops glycerin	food coloring

Procedure

1. Remove bread crusts and crumble bread into small pieces in bowl.
2. Add white glue, glycerin, and white vinegar.
3. Mix together until dough is supple and smooth.
4. Add food coloring and blend into the dough.
5. Make small objects and dry. It can take 2 days for drying.

■ Frozen Bread Molding Material
[Makes enough for 4 children]

NOTE: This dough can be expensive. Watch for sales of frozen bread dough. Baking produces hard-finished items that will not be eaten.

Materials

1 loaf frozen bread dough	nonstick cooking spray
1 egg white	small mixing bowl
2 teaspoons water	fork
cookie sheet	brush

Procedure

1. Defrost bread dough the day before using.
2. Break dough into smaller portions and shape as desired.
3. Place creations on cookie sheet sprayed with nonstick cooking spray.
4. Let rise for about 1 hour.
5. Beat egg white and water together in the small mixing bowl with the fork.
6. Brush egg white on creations.
7. Bake at 350°F for about 15–20 minutes.

■ Whole Wheat Flour Dough 1 (Flour and Salt)
[Makes 4–4 1/2 cups]

NOTE: This dough produces a light brown, speckled product that is very appealing to children. Whole wheat flour, which contains more oils than all-purpose flour, can go rancid rather quickly. Therefore, whole wheat flour should be stored in the freezer.

Materials

2 1/4 cups boiling water   mixing bowl
2 cups salt   kneading surface with extra flour
3 cups all-purpose flour   old cookie sheet
1 cup whole wheat flour

Procedure

1. Combine boiling water and salt in mixing bowl. Stir until salt is dissolved.
2. Slowly add flours, stirring after every addition. Make a stiff dough.
3. Knead on a floured surface. Fashion into objects.
4. Bake objects at 275°F for about 2 hours.
5. Objects can also air dry over several days.

■ Whole Wheat Flour Dough 2 (Flour and Vegetable Oil)
[Makes 5 cups, enough for 4 children]

NOTE: Children should keep their hands wet when they mix this clay. The moisture keeps the dough from sticking to their hands.

Materials

1 cup salt   mixing bowl
1 1/2 cups water   mixing spoon
2 tablespoons vegetable oil   cookie sheet
5 cups whole wheat flour

Procedure

1. Combine salt and water in bowl.
2. Stir in oil and flour.
3. Form desired shapes.
4. Place on baking sheet and bake at 325°F for 1 hour.

■ Dryer Lint Clay
[Makes 4 cups]

NOTE: Children can collect dryer lint for a couple of weeks before the project. Commercial laundries may be able to provide lots of lint. Oil of wintergreen keeps clay from becoming moldy. However, finished pieces will dry.

Materials

3 cups dryer lint	newspaper
2 cups water	pot
2/3 cup all-purpose flour	stove or heating element
3 drops oil of wintergreen (for preservative)*	mixing spoon

*Oil of wintergreen should not be eaten. If young children use this material, consider leaving the oil out.

Procedure

1. Combine lint and water in pot.
2. Stir in flour and oil of wintergreen.
3. Stir over low heat until mixture is stiff.
4. Take it out of the pot and place it onto newspaper to cool.
5. Fashion into objects. The objects will dry within a day or two.

■ Tissue Paper Bead Clay
[Makes 1 3/4 cups—enough for 2 children]

NOTE: The tissue paper gives the beads a delicate look. The clay is easy to work with, and the beads can be strung and attached to book marks. Water causes the tissue paper dyes to run. Therefore, use only one color of tissue paper in each batch.

Materials

1 1/2 cups shredded tissue paper	mixing spoon
1 cup boiling water	kneading surface
1/2 cup all-purpose flour	plastic straws
mixing bowl	paints and paintbrushes

Procedure

1. Combine tissue paper and boiling water in mixing bowl. Let stand for several hours.
2. Drain water.
3. Stir in flour bit by bit and knead clay.
4. Mold clay around straws to make beads.
5. Let beads air dry.
6. Remove straws and paint beads.

■ Clay Beads
[Makes about 2 cups]

NOTE: This clay is easy to work with. Interesting patterns can be created. These beads do dry.

Materials

2 cups baking soda
1 cup cornstarch
1 1/4 cups cold water
pot
stove or heating element
mixing spoon

small containers
food coloring
plastic wrap
drinking straws
wax paper

Procedure

1. Combine cornstarch and baking soda in pot.
2. Stir in water.
3. Heat mixture at medium heat, stirring constantly, until dough forms. This takes about 10–15 minutes.
4. Take pot off heating element and allow clay to cool.
5. Divide into sections and place each portion in a separate container. Add a few drops of food coloring to each section.
6. Cover each container with plastic wrap until you are ready to use it.
7. Break off pieces and roll into beads.
8. Use drinking straws to make holes through the beads.
9. Place beads on the wax paper to dry. It may take overnight to make them very hard.

■ Flower Petal Clay Beads
[Makes about 1 1/2 cups]

NOTE: Six cups of flower petals is quite a bit. Children could collect flower petals during late summer. Floral shops may provide discarded blossoms. The beads, of course, dry.

Materials

6 cups flower petals	mixing bowl
2/3 cup all-purpose flour	mixing spoon
2 tablespoons salt	toothpicks
1/3 cup water	wax paper

Procedure

1. Crumble and tear flower petals into small pieces. They will soften and smell wonderful.
2. Combine flour, salt, and water in mixing bowl.
3. Work in flower petals until the dough is soft and fun to work with.
4. Tear off small pieces of dough and roll into balls.
5. Use toothpicks to make holes through the beads.
6. Place beads on the wax paper. Let them air dry for a day or two.

■ Sawdust Clay 1 (Sawdust and Plaster of Paris)
[Makes 2 1/2 cups]

NOTE: Lumberyards will donate lots of sawdust. The trip to collect the sawdust is even interesting. No smoking is allowed near the sawdust, because small bits fill the air and it is easily combustible. Sawdust creations do dry.

Materials

2 cups sawdust	water to moisten
1 cup wallpaper paste	old dishpan
1/4 cup plaster of Paris*	wax paper

*Plaster of Paris may be hazardous to health if used extensively without protection. Children should perhaps use plaster of Paris outside. If used inside, children should perhaps wear goggles and dust filter masks.

Procedure

1. Combine sawdust, wallpaper paste, and plaster of Paris in the old dishpan.
2. Add enough water to make stiff dough.
3. Create objects and let air dry on wax paper.

■ Sawdust Clay 2 (Sawdust and Paste)
[Makes 1 cup]

NOTE: Finished products can be sanded.

Materials

1 cup sawdust	newspaper
1 cup thin paste or paper paste	old mixing bowl
food coloring (optional)	old cookie sheet

Procedure

1. Sawdust can be dyed by adding some food coloring to sawdust. Spread sawdust on newspaper to dry.
2. Combine sawdust and paste in old mixing bowl.
3. Fashion into objects.
4. Let air dry for 2 or 3 days or bake on old cookie sheet at 200°F for 1 to 2 hours.

■ Sawdust Clay 3 (Sawdust and Flour)

NOTE: Sift through sawdust first to take out large pieces. This clay produces hard, finished items.

Materials

2 cups sawdust   old dishpan
1 cup all-purpose flour   wax paper
water

Procedure

1. Combine sawdust and flour in old dishpan.
2. Add enough water to make stiff dough.
3. Fashion mixture into objects and place on wax paper.
4. Dry materials in the sun.

■ Glue–Dextrin Dough

[Makes enough dough to make 3 small pieces]

NOTE: Dextrin is found in artificial sweeteners. It can be purchased at a grocery store. Finished items will dry.

Materials

1 teaspoon white glue   palette knife or putty knife
1 cup powdered dextrin   old plate
1 tablespoon hot water   small bowl
1/4 cup patching plaster, whiting, disposable spoon
    Bon Ami, unscented talcum or   wax paper
    powdered chalk

Procedure

1. Pour white glue on plate.
2. Combine dextrin and water in small bowl.
2. Add dextrin solution to glue and mix well.
3. Add patching plaster 1 tablespoon at a time. Mix with palette knife and keep adding plaster until no more can be absorbed.
4. Knead dough and shape into desired objects.
5. Allow objects to dry on wax paper.

■ Sweet Smelling Dough
[Makes 3 cups, enough for 2 or 3 projects]

NOTE: This dough has a bright, sparkly color, and it has a pleasant smell. Consider adding glitter or bits of plastic confetti. Items produced from this dough will dry.

Materials

2 1/4 cups all-purpose flour
1 cup salt
2 tablespoons unsweetened powdered
   drink mix
4 tablespoons vegetable oil

1 cup water
large mixing bowl
mixing spoon
kneading surface with extra flour
airtight container

Procedure

1. Combine flour, salt, and unsweetened powdered drink mix in the large mixing bowl.

2. Add vegetable oil and water.

3. Stir until mixture is stiff.

4. Place dough on floured kneading surface and knead for 2 minutes or until it is smooth and pliable.

5. Creations will air dry after a couple of days.

6. Store any unused dough in airtight container.

■ Cotton Ball Dough
[Makes about 3 cups]

NOTE: Cotton ball dough makes great snowmen and winter dioramas. Items will dry.

Materials

3 cups cotton balls	pot
2 cups water	stove or heating element
1 cup all-purpose flour	mixing spoon
food coloring	paper towels

Procedure

1. Shred cotton balls into small pieces.
2. Pour cotton ball pieces into pot and add water.
3. Little by little add flour, stirring all the time.
4. Cook over low heat for about 5 minutes until mixture becomes stiff.
5. Remove dough from the pot and place on paper towels to cool.
6. Mold when cool.
7. Shapes will harden within a day.

■ Construction Paper Modeling Compound
[Makes about 3 cups]

NOTE: This mixture can go sour if left out too long.

Materials

2 cups colored construction paper, torn into small bits	blender
	mixing bowl
4 cups water	mixing spoon
1/2 cup all-purpose flour	kneading surface with extra flour

Procedure

1. Combine construction paper bits and 3 1/2 cups water in the blender. Blend about 30 seconds until a pulp forms.
2. Remove excess water.
3. Combine 1/2 cup water and flour in the mixing bowl.
4. Gradually add paper pulp to flour/water mixture.
5. Knead until it forms a stiff clay.
6. Finished products will dry within a day or two.

■ Coffee Clay
[Makes about 2 1/2 cups]

NOTE: Coffee clay has a nice smell and an interesting color. Baked items will become hard.

Materials

1/4 cup instant coffee powder   1/2 cup salt
3/4 cup warm water   mixing bowl
2 cups all-purpose flour   mixing spoon

Procedure

1. Dissolve coffee powder in a small portion of water.
2. Add flour and salt to the mixture.
3. Add rest of the water and stir until the dough is pliable.
4. Bake products at 300°F for 30 minutes or until hard.

■ Coffee Grounds Clay
[Makes 3 cups—enough for 2 children]

NOTE: Coffee grounds clay will not dry hard. Although it looks like ordinary clay, it will crumble if manipulated too much. It makes nice (but temporary) coffee grounds castles.

Materials

1 cup dry coffee grounds (fresh   water
    or used)   mixing bowl
1 cup cornmeal   mixing spoon
1/4 cup salt

Procedure

1. Combine dry ingredients.
2. Add enough water to make dough.
3. Model clay and have fun!

■ Coffee Grounds Rocks
[Makes about 2 1/2 cups, enough for about 10 rocks]

NOTE: Collect coffee grounds from the faculty room coffee pot at the end of the day. Leave them out to dry overnight.

Materials

1 cup coffee grounds	mixing bowl
1 cup all-purpose flour	mixing spoon
1/2 cup sand	wax paper
1/2 cup salt	cookie sheet
1 cup water	

Procedure

1. Combine coffee grounds, flour, sand, and salt in mixing bowl.
2. Add enough water to make a stiff dough.
3. Scoop out pieces of dough the size of golf balls and place on wax paper. They will air dry within 3 days.
4. Or bake on cookie sheet at 150°F for about 25 minutes.

■ Crepe Paper Molding Material
[Makes about 1 cup]

NOTE: Colored crepe paper pulp may stain hands and clothing. Consider adding several drops of vanilla extract or other flavoring to mask the rather unpleasant smell.

Materials

1 roll crepe paper	1/3 cup salt
water	mixing bowl
old bucket	mixing spoon
1/2 cup all-purpose flour	

Procedure

1. Shred crepe paper into small bits and place them in old bucket.
2. Cover crepe paper with water and let mixture soak overnight.
3. Thoroughly remove water and transfer one cup of pulp to mixing bowl.
4. Add flour and salt and stir until ingredients are thoroughly mixed.
5. Mold as if it were papier-mâché.

■ Sand Clay
[Makes 2 1/2 cups—enough for 2 children]

NOTE: When dry, projects resemble rocks, hard and grainy.

Materials
1 1/2 cups sand	powdered tempera paint (optional)
1 teaspoon alum	stove or heating element
1/2 cup cornstarch	pot
3/4 cup boiling water	mixing spoon

Procedure
1. Combine sand, alum, and cornstarch in pot.
2. Add boiling water and powdered tempera paint.
3. Heat at medium temperature until thickened.
4. Cool and model.
5. Air dry for several days.

■ Permanent Sand Clay
[Makes about 2 1/2 cups]

NOTE: Children like this clay's texture. Could this be done outdoors?

Materials
2 cups sand	stove or heating element
2/3 cup cornstarch	mixing spoon
1 1/2 cups liquid starch	kneading surface covered with wax
old pot	paper.

Procedure
1. Pour sand and cornstarch into old pot.
2. Mix in liquid starch.
3. Stirring constantly, cook over medium heat until a dough forms.
4. Remove from heat and let cool.
5. Knead mixture for about one minute before molding.

■ Sand Beads
[Makes 1 cup]

NOTE: Children like the grainy texture. It is messy at first, but it is worth the time. The beads do dry.

Materials

1 cup sand
1/4 cup white glue
disposable container
disposable mixing spoon

several smaller containers
tempera paints
toothpicks
wax paper

Procedure

1. Combine sand and white glue in disposable container.
2. Divide mixture between several small containers and add tempera paints if desired.
3. Pinch off bits of clay to make beads.
4. Use toothpicks to make holes for beads.
5. Dry overnight at least on wax paper.

■ Vermiculite Carving Compound
[Makes about 4 cups]

NOTE: Vermiculite, a soil additive, can be bought at hardware stores and plant nurseries. Vermiculite carving compound is a new experience for many children. This compound produces hard finished items.

Materials

2 cups plaster of Paris*
1 cup water
3 cups vermiculite
old bucket

mixing spoon
half-gallon wax-carton milk
  container, empty and clean
carving tools

*Plaster of Paris may be hazardous to health if used extensively without protection. Children should perhaps use plaster of Paris outside. If used inside, children should perhaps wear goggles and dust filter masks.

Procedure

1. Combine plaster of Paris and water in old bucket.
2. Slowly add vermiculite and combine thoroughly
3. Pour into container and let harden for at least an hour. The harder the compound, the harder the ability to carve.
4. Peel away container and carve.

■ Cinnamon Dough
[Makes about 1 1/2 cups]

NOTE: The smell is wonderful! It does not easily dry out. Consider making Christmas items from the material. The final products are not edible.

Materials

1 cup all-purpose flour
1/2 cup salt
2 teaspoons cream of tartar
2 teaspoons vegetable oil
1 cup water
about 6 drops red food coloring
about 6 drops green food coloring
2 tablespoons cinnamon

2 tablespoons allspice
2 mixing bowls
2 mixing spoons
old pot
stove or heating element
kneading surface with extra flour
airtight container

Procedure

1. Mix flour, salt, and cream of tartar in a bowl.
2. Stir in cinnamon and allspice.
3. In the other bowl, add the food colorings to the water. Red and green should form brown.
4. Add water mixture and vegetable oil to dry ingredients and stir.
5. Pour into old pot and cook mixture for about 3 minutes, stirring constantly.
6. Remove dough from pot and knead until it is pliable and smooth.
7. Allow to cool.
8. Shape as desired and let air dry.
9. Store any unused dough in airtight container.

■ Ice Sculptures
[Makes enough for 10 sculptures]

NOTE: These make great projects for cold winter mornings and hot summer afternoons.

Materials

10 disposable aluminum pie pans	water
10 cups of clay	food coloring (optional)

Procedure

1. Line pie pan with clay to form a creative mold.
2. Fill mold with water and add food coloring if desired.
3. Place in a freezer.
4. The next day take the pans from the freezer and remove clay.
5. Place the sculptures outside if it is cold enough.

■ Real Clay
[Makes about 5 cups]

NOTE: Real clay is often found along riverbanks or washed-out areas. Squeeze together what you think might be clay. If it sticks together when you release your hand, it probably is clay. While real clay products can be placed in a kiln, they can also be just left to dry. Real clay can also be purchased in craft stores.

Materials

about 6 cups clay	old bowl
bucket	cloth big enough to line bowl
3 layers newspaper	water
hammer	flat surface
old sieve	airtight container

Procedure

1. Fill bucket with clay and be prepared that the clay will be heavy.
2. Pour clay onto layers of newspaper and let it dry thoroughly.
3. Remove any debris such as rocks or twigs.
4. Use hammer to break up any small masses of clay.
5. Sift dry clay through old sieve to remove small pieces of debris.
6. Pour clay back into bucket and cover with water. Let mixture sit for about a day.
7. Discard any extra water. Pour clay into old bowl lined with cloth.
8. Allow clay to dry enough so that it is easy to work with.
9. Now wedge clay to get out any air bubbles. To wedge clay, pick it up and throw it on the work surface until no more air bubbles are present.
10. Store clay in airtight container.

2
Salt Map Mixtures

Salt map mixtures have been around for generations. Low-tech and low-cost, these doughs have been used by children for many, many years. These recipes have other uses than salt maps. For example, these recipes, while the dough is soft, could be used in activities such as "Show me a . . ." and the adult provides the example. "Show me" a plateau, or a valley, or a delta. Beyond geography, children could make examples of biomes, or molecular structure, or cell structure. Children enjoy making dioramas, and the salt map doughs would be much better than glues or clays.

The differences in salt map mixtures are all about the ratio of salt to flour. These maps are best made when humidity levels are low. The higher the humidity, the more moisture the salt will retain. Also, children should add water very slowly and in small amounts to the dry ingredients. Somehow the mixture can go from dough to soup with only a few extra drops of water.

■ Salt Map Mixture 1 (Smooth Texture, Short-Term Stability)
[Makes 2 1/2 cups—enough for 1 map]

NOTE: This mixture is not as grainy as Salt Map Mixtures 3 and 4. It can be shaped and reshaped while damp. This mixture is preferable for maps that are to be used on a short-term basis because the dried shapes may crumble over a long period of time.

Materials

1 cup salt	mixing spoon
1 cup all-purpose flour	wood or cardboard base
approximately 1 cup water	paints and paintbrushes
mixing bowl	

Procedure

1. Mix salt and flour in bowl. Add enough water, slowly and in small amounts, to make soft dough.
2. Apply mixture to base in layers. Allow several hours drying time before adding each successive layer.
3. Build up higher elevations by applying new layers.
4. Paint when salt map is thoroughly dry (1–3 days).

■ Salt Map Mixture 2 (Clay-Like, Fast Drying)
[Makes 3 1/2 cups—enough for 1 map]

NOTE: Alum, found in the spice section of a grocery store, makes this mixture dry more quickly than other mixtures in this chapter. The dough is not grainy and can be rolled to make coils. The coils can then be shaped to resemble mountains. Edges can be smoothed out. Thus, the mountains are hollow, and they dry quickly. Paints can be added during the mixing instead of after the map has dried.

Materials

1 cup salt	mixing bowl
2 cups all-purpose flour	mixing spoon
2 teaspoons alum	wood or cardboard base
approximately 1 cup water	paints and paintbrushes

Procedure

1. Mix salt, flour, and alum together in the bowl.
2. Add enough water, slowly and in small amounts, to make stiff dough.
3. Form some dough into coils and apply other dough to base.
4. Smooth out coils to form mountains, river banks, or other features.
5. Paint when thoroughly dry (1–2 days).

■ Salt Map Mixture 3 (Grainy Texture, Long Lasting)
[Makes 3 1/2 cups—enough for 2 maps]

NOTE: This grainy, white mixture resembles earth's rough texture. Because the final product is quite durable, this recipe should be used on any salt maps that will be kept for a long time.

Materials

2 cups salt	mixing spoon
1 cup all-purpose flour	wood or cardboard base
approximately 1 1/2 cups water	paints and paintbrushes
mixing bowl	

Procedure

1. Mix salt and flour in bowl. Add enough water, slowly and in small amounts, to make soft dough.
2. Apply immediately to base.
3. Paint when salt map is thoroughly dry (1–2 days).

■ Salt Map Mixture 4 (Great Texture, Slow Drying)

[Makes 3 1/2 cups—enough for 2 maps]

NOTE: This formula has a lumpy texture, similar to that of the earth's surface. Therefore, maps made from this recipe look terrific. However, this mixture takes longer to dry than other recipes.

Materials

1 1/2 cups coarse salt, such as that found on soft pretzels	mixing bowl
	mixing spoon
1 cup all-purpose flour	wood or cardboard base
approximately 1 1/2 cups water	paints and paintbrushes

Procedure

1. Mix coarse salt and flour in bowl.
2. Add enough water, slowly and in small amounts, to make soft dough.
3. Apply mixture to base immediately.
4. Paint when salt map is thoroughly dry (2–3 days).

■ Salt Map Mixture 5 (Colorful, Fragrant)

[Makes about 2 cups of 1 color]

NOTE: This salt map dough contains color, so children will not need to paint it. The nice smell makes it fun to mold.

Materials

1 small box fruit-flavored gelatin	mixing bowl
1/2 cup salt	mixing spoon
2 cups all-purpose flour	wood or cardboard base
approximately 3/4 cup water	

Procedure

1. Combine gelatin, salt, and flour in mixing bowl.
2. Add a small amount of water to mixture and combine.
3. Continue to add small amounts of water until mixture resembles clay.
4. Mold and allow to dry on base.

3
Papier-Mâché

The term *papier-mâché* means "chewed paper." Its origin dates back to ancient China. The French popularized this moldable pulp in Europe during the 1600s. Materials are inexpensive and easy to obtain. Because papier-mâché pulp can develop mold over time, you can add some salt to act as a preservative. Papier-mâché paste can also be refrigerated.

Papier-mâché can clog drains. Drain water from any remaining papier-mâché and toss unused pulp into a waste basket.

■ Papier-Mâché Pulp
[Makes 1 quart pulp—enough for 3 children]

NOTE: This recipe, like most of the recipes in this chapter, takes more than one day.

Materials

16 pages of newspaper	bucket
water	blender
2 tablespoons ground chalk or spackle (adds density and improves color)	stove or heating element
	large pot
1 tablespoon linseed oil (adds sturdiness and malleability)	sieve
	mixing bowl
1 tablespoon salt (serves as preservative)	
2 tablespoons papier-mâché paste (see p. 53)	

Procedure

1. Tear newspaper into 1-inch squares.
2. Place newspaper pieces in the bucket. Cover with water and soak overnight.
3. The following day, drain most of water from bucket. Move paper to large pot with 1/2 gallon water and boil for 20 minutes.
4. Pour portions of boiled newspaper into blender and pulse until paper becomes pulp. Pour pulp into bucket and blend remaining batches.
5. Strain pulp through sieve to remove extra water.
6. Form a ball of pulp and squeeze more water (but not all water) from pulp.
7. Transfer pulp to mixing bowl. Add rest of ingredients and mix.
8. Pulp can now be molded, sculpted, carved, and so on. When it is dry, it can be sanded and painted.

■ Molding with Papier-Mâché

NOTE: Papier-mâché strips should be layered over an existing model. When the papier-mâché dries, the original mold is removed. Plastic bowls and boxes make the best molds because they will neither break nor absorb moisture from the papier-mâché.

Materials

mold (plastic bowl, box, etc.)
petroleum jelly
newspaper
container filled with water
papier-mâché paste (see p. 53)

white glue or gesso (to use as a sealer) and paintbrush

Procedure

1. Cover work area with some newspaper.
2. Apply coating of petroleum jelly all over mold, so that the mold can be removed when the project is done. Remember to cover lip or edges.
3. Rip other newspaper into strips.
4. Dip first layer of newspaper into only plain water.
5. Get rid of excess water and place strips on mold. Completely cover mold. Allow to dry.
6. For all subsequent layers, dip each strip of newspaper into papier-mâché paste before applying it to previous layer.
7. Allow each layer to dry before adding the next layer.
8. After layers are dry, remove mold.
9. Seal with white glue or gesso.
10. When sealer is dry, paint.

■ Very Soft Papier-Mâché

NOTE: This soft papier-mâché allows children to do fine work, such as facial features, on big papier-mâché pieces. This recipe is great for Halloween projects.

Materials

Soft papers such as paper towels, napkins, tissues, or toilet paper

white glue

Procedure

1. Wad soft papers and dip them into white glue.
2. Make into desired shapes.

■ Making Large Papier-Mâché Objects

NOTE: One advantage that papier-mâché has over clay is that the former can be used to make large but light-weight items. Children can make just about anything they can imagine using papier-mâché.

Materials

full sheets of newspaper and strips of newspaper
coat-hanger wire (or chicken wire for very large objects)
papier-mâché paste (see p. 53)

masking tape
white glue (to use as a sealer) and paintbrush
paints and paintbrushes

Procedure

1. Cover work area with some newspaper.
2. Make frame for object by forming wadded newspaper sheets into approximate size and shape of desired object.
3. Add wire to frame to help retain its shape.
4. Wrap masking tape around wire and newspaper balls.
5. Paint shape with papier-mâché paste.
6. Cover with strips of newspaper. Make sure that all masking tape is covered with newspaper strips. Allow shape to dry.
7. Apply second coat of papier-mâché paste. Cover again with small pieces of newspaper strips and allow to dry.
8. Seal with white glue and allow shape to dry.
9. Paint object.

■ Papier-Mâché Dough
[Makes 4 cups—enough for 3 children]

NOTE: Ordinarily, papier-mâché is applied to other surfaces (e.g., balloons or picture frames). In this recipe, papier-mâché dough itself can be shaped into small objects.

Materials

newspaper
water
1 cup all-purpose flour
1/2 cup salt

bucket
blender
paints and paintbrushes

Procedure

1. Tear newspaper into 1-inch squares. Place in bucket.
2. Cover with water and soak until the next day.
3. Place 1 cup soaked paper and water into blender. Blend until pulp forms.
4. Strain out water and place pulp in mixing bowl. Continue with remaining newspaper bits until 3 cups of pulp are produced.
5. Combine pulp with flour and salt to form a dough.
6. Form objects. An object with more bulk than a grapefruit should be made using a frame of newspaper, wire, and masking tape.
7. Let dry and then paint.

■ Papier-Mâché Maracas

NOTE: Maracas, usually played in pairs, may have been invented in Puerto Rico as far back as the year 1500. Maracas are often associated with Latin American music. Here children have a chance to link art and music. The patterns they paint on their maracas are most interesting.

Materials

cardboard tubes from rolls of toilet paper or paper towels
scissors
small balloons
masking tape

newspaper
papier-mâché paste (see p. 53)
dry lima beans
paints and paintbrushes
acrylic gloss and brush (optional)

Procedure

1. Cover work area with a layer of newspaper.
2. Cut 2 cardboard tubes into 3-inch lengths.
3. Blow up 2 balloons to size of large lemons. Tie ends.
4. Tape each balloon to end of cardboard tube. This is the frame of the maraca.
5. Cut newspapers into small strips (about 6 inches by 2 inches).
6. Dip newspaper strips into papier-mâché paste, and apply to maraca form.
7. Cover entire form of maraca, except the very tip of balloon.
8. Let dry and apply second coat of newspaper strips and papier-mâché paste.
9. Let second layer dry.
10. Pop balloon. Insert several dry lima beans.
11. Cover hole with newspaper strips and papier-mâché paste. Let dry.
12. Paint and seal with acrylic gloss, if desired.

■ Papier-Mâché Piñata

NOTE: A piñata is a good finale to a study of Mexico. People also have piñatas at birthday parties. Note that it takes more than a week to make.

Materials

balloon	paints and paintbrushes
string	feathers, glitter, and other colorful
newspaper	and decorative materials
papier-mâché paste (see p. 53)	small pieces of candy and small toys
white glue	for piñata "stuffing"

Procedure

1. Blow up balloon and knot it. Tie a piece of string around knot.
2. Rip newspaper into 3-inch squares.
3. Dip a piece of newspaper into papier-mâché paste and apply newspaper to balloon.
4. Repeat step 3 until balloon is entirely covered—except for a small area near the knot of the balloon, which will serve as an opening for stuffing the piñata (leave the opening large enough to accommodate candy and toys).
5. Hang balloon in a doorway overnight to dry.
6. Apply a second layer of newspaper squares the next day, and let dry overnight.
7. Repeat process until 6 layers have been applied. Let project dry.
8. Attach other features (such as cones to make a star) with papier-mâché paste.
9. Pop balloon and remove if possible.
10. Fill piñata cavity with candy and toys.
11. Apply one or more layers of papier-mâché over hole. Let dry.
12. Make several holes for a string to hang the piñata when done.
13. Seal with white glue.
14. After glue is dry, paint and add feathers, glitter, and other colorful and decorative materials.

4
Plaster of Paris

Plaster of Paris is a powdered form of gypsum. When water is added to plaster of Paris, the gypsum reforms and hardens. Heat is a by-product of this process.

Plaster of Paris projects are probably best done outside. Plaster of Paris may be hazardous to health if used extensively without protection. The dust can damage ears, eyes, lungs, and skin if precautions are not taken. Wear safety goggles and a dust filter mask when mixing plaster of Paris. Some experts suggest mixing plaster of Paris when children are not in the room. Use rubber gloves whenever possible. Also, plaster of Paris gives off heat as it solidifies. Although it could probably not burn a child, remind everyone to practice safety first.

Plaster of Paris should not be poured down the sink because it will clog drains. Clean all utensils as soon as possible so that the plaster will not harden on them. Whenever possible, use plastic containers that flex slightly. The plaster of Paris will crack and fall off when the flexible container is bowed.

■ Plaster of Paris Fossil
[Makes 4 small fossils]

NOTE: A fossil of a bone is usually not the real bone. Millions of years ago an animal fell into a clay-like substance and died. Over time the bone rotted away, leaving an impression. Water containing dissolved minerals entered the impression over thousands of years. Later, the water seeped out and the minerals remained, filling the impression. Eventually the minerals hardened into the shape of the original bone.

Materials

wax paper
1 cup soft clay that will not harden overnight
item to make a fossil imprint (e.g., a clean bone or twig)

petroleum jelly
1 cup plaster of Paris
approximately 1/2 cup water
disposable mixing bowl
disposable mixing spoon

Procedure

1. Divide clay into 4 sections and place on wax paper. Make sure clay is soft and pliable.

2. Coat bone, stick, or other object with petroleum jelly. Press object into clay and then carefully remove it. Make sure it has left a good imprint. Repeat with other 3 portions of clay.

3. Pour plaster of Paris into disposable bowl.

4. Add enough water, slowly and in small amounts, to make the plaster of Paris creamy.

5. Pour plaster of Paris over clay and into imprints. Let plaster harden overnight.

6. The next day, peel off the clay. The fossil is what remains.

■ Plaster of Paris Dinosaur Eggs or Bird Eggs
[Makes 4 eggs]

NOTE: These "eggs" can be added to dioramas. Children can research a specific animal's egg size and coloration.

Materials

1 cup plaster of Paris
approximately 1/2 cup water
disposable mixing bowl
disposable mixing spoon
funnel

4 small, oval balloons
4 pieces of string
large bucket, half filled with water
sandpaper
paints and paintbrushes

Procedure

1. Mix plaster of Paris and water in mixing bowl. The mixture should be smooth and creamy.
2. Place funnel in opening of 1 balloon.
3. Spoon some plaster of Paris mixture through funnel into balloon. The balloon will expand as more plaster is added. The plaster will take the shape of the balloon. Do not stretch balloon to its limits.
4. Pinch balloon opening before removing funnel.
5. Tie balloon opening with string.
6. Float balloon in water bucket so that there will not be a flat side.
7. Repeat process with rest of balloons.
8. Leave plaster-filled balloons in bucket for several hours.
9. Remove balloons and dry.
10. Cut away balloons.
11. If necessary, sand off the points where balloons were tied.
12. Let "eggs" age for a day or 2 and then paint.

■ Plaster of Paris Dinosaur Footprints
[Makes 4 footprints]

NOTE: A fossil can be broadly defined as a hardened trace of an animal or plant. A dinosaur's footprint is a type of fossil.

Materials

petroleum jelly
plastic replica of a dinosaur
1 cup plaster of Paris
approximately 1/2 cup water

disposable mixing bowl
disposable mixing spoon
wax paper

Procedure

1. Coat dinosaur replica's feet with petroleum jelly. The petroleum jelly will keep the plaster of Paris from sticking to the dinosaur replica.
2. Pour plaster of Paris into disposable bowl.
3. Add enough water, slowly and in small amounts, to make the plaster of Paris creamy.
4. Spoon the plaster of Paris on wax paper.
5. Wait until plaster of Paris is fairly stiff.
6. Have dinosaur replica "step" into and out of plaster of Paris. The replica should have left footprints.
7. Let plaster of Paris dry overnight. The next day examine the footprints.

■ Plaster of Paris Plant Imprints
[Makes 3 imprints]

NOTE: Paleontologists have found many fossilized plant imprints. This formula replicates the process.

Materials

fresh but sturdy leaves or flowers   disposable mixing bowl
1 cup plaster of Paris   disposable mixing spoon
approximately 1/2 cup water   wax paper

Procedure

1. Pour plaster of Paris into disposable bowl.
2. Add enough water, slowly and in small amounts, to make the plaster of Paris creamy.
3. Spoon plaster of Paris onto wax paper.
4. Wait until plaster of Paris is fairly stiff.
5. Press fresh leaf or flower into plaster. Carefully remove. An imprint in the plaster should remain.
6. Let plaster of Paris dry overnight. The next day examine the imprint.

■ Fresco
[Makes 1 fresco—enough for 1 child]

NOTE: A fresco is an artwork that is painted into the plaster on a wall. It becomes part of the wall and cannot be removed like a painting on canvas. First, the artist draws the work on paper. The paper is called a cartoon. Then a layer of plaster is applied to the wall. The cartoon is laid against the plaster and traced. Next, the artist applies the paints onto the wet plaster. When the plaster dries, the pigments become part of the plaster and thus part of the wall. Michelangelo and Diego Rivera are among the artists who became masters of the fresco technique.

Materials

1 cup plaster of Paris
approximately 1/2 cup water
disposable mixing bowl
disposable mixing spoon
disposable aluminum pie pan

piece of paper size of aluminum
  pie pan
toothpicks
tempera paints

Procedure

1. Mix plaster of Paris and water together in disposable bowl. Mixture should be smooth and creamy.
2. Pour mixture into aluminum pie pan.
3. Let mixture harden somewhat.
4. While mixture hardens, draw a cartoon of final product on paper.
5. Using a toothpick, outline design into plaster.
6. Using another toothpick, mix some of the tempera paint into the plaster. The plaster and paint are now one.
7. Using new toothpicks for each new color, complete the fresco.
8. Allow it to harden completely and then remove it from aluminum pie pan.

■ Bas Relief
[Makes 1]

NOTE: A bas relief is a type of sculpture that is 3-dimensional on one side and flat on the other. The relief can be hung on a wall. Almost all ancient cultures produced bas reliefs.

Materials

2 cups nonhardening clay
petroleum jelly
2 cups plaster of Paris
approximately 1 cup water
disposable aluminum tray

disposable mixing bowl
disposable mixing spoon
items to make an impression, such as
    seashells, twigs, small toys
soft cloth

Procedure

1. Spread nonhardening clay in disposable aluminum tray.
2. Cover seashells or other items with petroleum jelly. Press items into clay and then remove.
3. Combine plaster of Paris and water in disposable bowl.
4. Pour plaster of Paris into clay and impressions.
5. Let plaster of Paris harden overnight. Remove bas relief from clay.
6. Remove any bits of clay with soft cloth.

■ Mosaics
[Makes 5 mosaics]

NOTE: A mosaic is an art form dating back to ancient Mesopotamia. Mosaic artifacts have been discovered in many cultures, ranging from the Persians to the Maya. In a mosaic, small pieces of stone or other materials are embedded in mortar to form an image. In this project, the plaster of Paris serves as the mortar.

Materials

2 cups plaster of Paris
approximately 1 cup water
disposable mixing bowl
disposable mixing spoon

5 disposable aluminum pie pates
items to embed in mosaic (stones,
    shells, beans, broken crockery,
    smooth pieces of glass)

Procedure

1. Mix plaster of Paris and water together in disposable bowl. Mixture should be smooth and creamy.
2. Pour mixture into aluminum pie pan.
3. Let mixture harden somewhat.
4. While mixture hardens, plan a design for the mosaic.
5. When mixture is fairly stiff, embed items according to plan.
6. Let mosaics harden overnight. Remove from pans.

■ Faux Marble
[Makes 1 project]

NOTE: The glue strengthens the plaster of Paris. The tempera paint coloring allows for creativity.

Materials

1/2 cup water
1 tablespoon white glue
1 cup plaster of Paris
tempera paint
disposable mixing bowl

disposable mixing spoon
plaster of Paris mold, made of plastic
    or rubber (can be purchased at
    craft stores)

Procedure

1. Combine water and white glue in mixing bowl.
2. Add enough plaster of Paris to make a creamy and smooth mixture.
3. Pour tempera paint on top of plaster of Paris. With mixing spoon, swirl paint into plaster of Paris.
4. Pour mixture into mold and let it harden. Release from mold.

■ Sidewalk Chalk
[Makes 4 sticks chalk]

NOTE: This chalk is great for hopscotch and sidewalk pictures. Do not use it on chalkboards.

Materials

2 tablespoons powdered tem-
    pera paint
1 cup plaster of Paris
approximately 1/2 cup water

disposable mixing bowl
disposable mixing spoon
small paper cups

Procedure

1. Combine powdered tempera paint and plaster of Paris in disposable bowl.
2. Add water slowly and in small amounts until the plaster of Paris is creamy.
3. Spoon mixture into small paper cups.
4. Let mixture harden overnight.
5. Peel off paper and use chalk.

■ Volcano

[Makes 1 volcano]

NOTE: Follow the procedure below to make the volcano. See Chapter 16 for the procedure to "activate" the volcano.

Materials

small, empty, and clean plastic yogurt
   container
flat pan
small piece of aluminum foil to cover
   mouth of yogurt container

2 cups plaster of Paris
approximately 1 cup water
disposable mixing bowl
disposable mixing spoon
paints and paintbrushes

Procedure

1. Place empty yogurt container in middle of pan. Cover mouth of yogurt container with aluminum foil.

2. Combine plaster of Paris and water in disposable mixing bowl.

3. When plaster is fairly stiff, sculpt it over yogurt container. Scrape away any plaster that is directly on aluminum foil.

4. Using spoon, sculpt plaster so that it looks like the sides of a volcano. Let harden.

5. Cut away the aluminum foil to reveal the mouth of the container. The mouth of the container will be the mouth of the volcano.

6. Once plaster of Paris dries, paint volcano.

■ Plaster of Paris Draped Cloth
[Makes about 2 cups]

NOTE: This stiffened fabric makes excellent ghosts for Halloween or angels for Christmas. It can also cover chicken wire for very large projects. The cloth dries rapidly, because alum causes the plaster of Paris to harden quickly.

Materials

newspaper
2 cups plaster of Paris
1 teaspoon alum
approximately 1 cup water
fabric, sheets, or netting

disposable mixing bowl
disposable mixing spoon
bottle, inverted vase, or other object
   serving as a stand
paints and paint brushes (optional)

Procedure

1. Cover work area with newspaper.
2. Combine plaster of Paris and alum in mixing bowl.
3. Add enough water, slowly and in small amounts, to make plaster a thick consistency.
4. Dip fabric in plaster of Paris.
5. Remove fabric and drape over stand. Arrange fabric to suit taste.
6. Once the fabric stiffens, do not try to rearrange it.
7. Paint fabric when dry.

5
Glues and Pastes

What is the difference between a glue and a paste? I think a glue can be poured or squeezed on. I think a paste has to be applied with a brush or other tool. Both glues and pastes demonstrate cohesion and adhesion. The glue has to stick to itself (cohesion), and it has to stick to the surfaces it is gluing (adhesion).

■ Homemade Glue
[Makes about 1 cup]

NOTE: This glue is very useful, but it must rest for at least 12 hours before it can be used.

Materials

2 tablespoons corn syrup	pot
2 teaspoons white vinegar	stove or heating element
3/4 cup water	2 mixing spoons
1/2 cup cornstarch	mixing bowl
3/4 cup very cold water	airtight container

Procedure

1. Combine corn syrup, vinegar, and 3/4 cup water in pot and heat until it begins to boil.
2. Combine cornstarch and very cold water in mixing bowl.
3. Slowly add cornstarch–water mixture to the mixture in pot.
4. Stir until thoroughly mixed.
5. Remove from heat and let rest overnight.
6. Store in an airtight container.

■ Homemade Paste
[Makes 1 pint]

NOTE: Homemade paste will last for 2 or 3 months without being refrigerated; it is a good paste to use with papier-mâché.

Materials

1/4 cup sugar	pot
1/2 cup flour	stove or heating element
1/2 teaspoon alum (available in grocery stores in spices section)	mixing spoon
	brush
1 3/4 cups water	airtight container
1 tablespoon salt (to act as a preservative)	

Procedure

1. Combine sugar, flour, and alum in pot.
2. Stir in 1 cup water.
3. Boil, stirring constantly, until mixture is clear.
4. Stir in 3/4 cup water and salt.
5. Pour into airtight container and let cool.
6. Apply with brush.

■ Papier-Mâché Paste
[Makes 2 quarts]

NOTE: Papier-mâché paste is obviously great for papier-mâché projects. Warm it slightly before using. It can also be thinned with water and used as a more traditional paste.

Materials

2 cups all-purpose flour	stove, heating element, or microwave
1/4 cup sugar	pot or microwave-safe container
2 quarts warm water	mixing spoon
1 tablespoon salt (to act as a preservative)	airtight storage container

Procedure

1. Combine flour and sugar in pot or microwave-safe container.
2. Add a bit of warm water to make a thick paste. Slowly add rest of water, stirring with each addition.
3. Boil, stirring continuously, until mixture is thick and clear. Or microwave mixture at high setting for 2 minutes, stir, and microwave again at high setting for another 2 minutes.
4. Cool and store in airtight container.

■ Decoupage and Collage Glue
[Makes 1 pint]

NOTE: Decoupage projects take several days. Each application of glue must thoroughly dry before another layer is added.

Materials

1 1/2 cups white glue	container
1/2 cup water	mixing spoon

Procedure

1. Combine white glue and water in container.
2. Apply 1 layer of glue for collage. Add several layers, with drying time between, for decoupage.

■ Emergency Paste
[Makes 1 pint]

NOTE: Make this paste when children want glue "now" and all the white glue has disappeared.

Materials

1/2 cup all-purpose flour	mixing spoon
3/4 cup cold water	stove or heating element
1 cup boiling water	airtight container
pot	

Procedure

1. Combine flour and cold water in pot.
2. Add boiling water.
3. Cook, stirring constantly, for 3 minutes.
4. Cool and store in airtight container.

■ Paper Paste
[Makes 1 pint]

NOTE: Papers glued with this paste will dry flat. Sometimes commercial white glues buckle paper.

Materials

9 ounces white dextrin (available at grocery stores in artificial sweeteners section)	1 teaspoon alum (available at grocery stores in spices section)
1 3/4 cups water	pot
2 teaspoons sugar	stove or heating element
2 teaspoons glycerin (available in small bottles in pharmacies)	mixing spoon
	candy thermometer
	airtight container

Procedure

1. Combine dextrin and water in pot. Heat, stirring constantly, to 140°F.
2. Add sugar, glycerin, and alum and heat until mixture becomes clear.
3. Let mixture cool slightly and then pour into airtight container. Let cool completely.

■ Nonpaper Glue
[Makes 1/2 cup]

NOTE: Use this glue in its liquid state to attach glass to glass or wood to wood. It will bond metal to metal when it is gelled. This glue does not always repair ceramics or pottery.

Materials

2 envelopes (2 tablespoons) unflavored gelatin
1 1/2 tablespoons water
3 tablespoons skim milk
mixing bowl

mixing spoon
pot
stove or heating element
airtight container

Procedure

1. Combine gelatin and water in mixing bowl.
2. In the pot, scald skim milk.
3. Pour milk into gelatin mixture.
4. Pour into airtight container and store in refrigerator.
5. Glue will gel in container. To liquefy, place container into a larger receptacle containing a bit of hot water.

■ Colored Glues

NOTE: Glue makers are now selling colored glues. These colored glues add spice to posters and other projects. Colored glues are easy to make and are usually cheaper than purchased glues. The following formula is especially great if only a small amount of one color is needed.

Materials

white glue in squeeze bottles   poster paint

Procedure

1. Open bottles of glue.
2. Add a bit of poster paint to each bottle. Replace lids and shake. The finished product will be darker colored than the original paint. Do not add too much paint, or the liquid will affect the glue's effectiveness.
3. For a variation, do not shake the bottle too much. The marbled glue will produce interesting results.

■ Glitter Glues

NOTE: Glitter glues, which produce puffy, interesting touches to projects. Homemade glitter glues are fun and cheap to make. Glitter, as always, can end up in unexpected places.

Materials

white glue in squeeze bottles   containers of glitter

Procedure

1. Open bottles of glue.
2. Add a bit of glitter to each bottle. Replace lids and shake.

■ Molding Paste
[Makes about 4 cups]

NOTE: This paste sort of turns paper into papier-mâché.

Materials

1 cup all-purpose flour   stove or heating element
5 tablespoons sugar   mixing spoon
4 cups water   airtight container
pot

Procedure

1. Combine ingredients in pot.
2. Heat, stirring constantly, until mixture boils.
3. Reduce heat. Keep stirring until mixture becomes thick.
4. Remove from heat and cool in airtight container.
5. Refrigerate any unused portion.

■ Quick Molding Paste
[Makes about 1 1/2 cups]

NOTE: This recipe requires no heat, and it is extremely easy to make.

Materials
1 cup all-purpose flour
1 cup water
mixing bowl

mixing spoon
airtight container

Procedure
1. Combine flour and water in mixing bowl until mixture is thick.
2. Pour into airtight container and store unused portion in the refrigerator.

■ Sugar Glue
[Makes about 2 cups]

NOTE: This glue is a good emergency glue when a project is due right away and the commercial glue is all gone.

Materials
1/2 cup sugar
1/2 cup all-purpose flour
1 1/2 cups water
pot

stove or heating element
mixing spoon
airtight container

Procedure
1. Combine flour, sugar, and water in pot.
2. Stirring constantly, bring mixture to a boil.
3. Reduce heat and cook until mixture is thick.
4. Pour into container and allow to cool. Add lid.
5. Store unused portion in the refrigerator.

■ Cornstarch Squeeze Glue
[Makes 2 cups]

NOTE: The cornstarch–water mixture can act either as a liquid or as a solid.

Materials

3 tablespoons cornstarch
4 tablespoons cold water
2 cups hot water
pot

stove or heating element
mixing spoon
clean squeeze bottle such as an old
    dish detergent bottle

Procedure

1. Combine cornstarch and cold water in pot.
2. Stir in hot water.
3. Heat at a medium temperature, stirring constantly, until mixture thickens.
4. Remove from heat and let cool.
5. Pour into squeeze bottle.
6. Refrigerate when it is not being used.

■ Casein Glue
[Makes about 3/4 cup]

NOTE: The ancient Egyptians made casein glue thousands of years ago.

Materials

1 quart skim milk
1 cup white vinegar
pot
stove or heating element
wooden spoon
cheesecloth
strainer

old construction paper
airtight container
disposable cup
1 1/2 tablespoons hot water
2 teaspoons borax
brush

Procedure

1. Combine skim milk and white vinegar in pot.

2. Heat mixture over low heat for about 15 minutes until film forms on top.

3. Place cheesecloth in strainer and pour liquid from pot into strainer.

4. Run water through strainer to remove any remaining vinegar.

5. Remove cheesecloth and casein glue (curd) from strainer and squeeze out any remaining liquid.

6. Remove curd and place between sheets of old construction paper for at least a day.

7. Remove curd and place in the airtight container.

8. To make casein glue, dissolve 2 teaspoons borax into 1 tablespoon of hot water in disposable cup.

9. Add 2 tablespoons of curd. Combine and let cool for a minute or two. Apply glue with a brush.

■ Sticker Gum
[Makes 1/4 cup]

NOTE: Children have fun with this recipe by making their own stickers.

Materials

1 tablespoon (1 envelope) unflavored gelatin
1 tablespoon sugar
1/4 cup boiling water
1/2 teaspoon flavoring (such as vanilla or strawberry)

small mixing bowl
spoon
airtight container

Procedure

1. Combine gelatin and boiling water in small mixing bowl.
2. Add sugar and flavoring.
3. Let cool for a few minutes and then pour into airtight container.
4. Store in refrigerator or it will mold.
5. To use, place container in warm water to liquefy gum. Paint surface with a bit of mixture.
6. Let dry. Moisten to activate the stickiness.

■ Sticker Making Solution
[Makes 1/4 cup]

NOTE: This recipe has a strong, pleasant smell.

Materials

1/4 cup hot water
3 tablespoons flavored gelatin
mixing bowl

spoon
old paintbrush
small cutouts from magazines

Procedure

1. Pour gelatin into bowl.
2. Add hot water and stir until thoroughly mixed.
3. Paint solution on backs of cutouts and allow to dry for about an hour.
4. Moisten back to activate.

■ Envelope Glue

[Makes about 1/2 cup]

NOTE: Children can make their own envelopes from a sheet of paper and this glue.

Materials

6 tablespoons white vinegar
4 tablespoons (4 envelopes) unfla-
    vored gelatin
1 tablespoon vanilla extract
pot

stove or heating element
mixing spoon
airtight container
old brush

Procedure

1. Pour white vinegar into pot and bring to a boil.

2. Dissolve gelatin in white vinegar.

3. Add extract to provide flavor.

4. Pour into airtight container.

5. When ready to use, melt small amount and apply to paper. Fold paper to make envelope.

■ Transparent Mending Glue

[Makes 1 1/2 pints]

NOTE: This glue repairs torn pages in books. To use, place a small amount of glue over tear and apply small piece of white tissue paper. The tissue paper will become transparent when glue dries.

Materials

3/4 cup rice flour
2 tablespoons sugar
3/4 cup water
2 1/2 cups hot water
1 tablespoon salt (to act as a
    preservative)

pot
stove or heating element
mixing spoon
airtight container

Procedure

1. Combine rice flour, sugar, and 3/4 cup water in pot.

2. Add 2 1/2 cups hot water.

3. Bring mixture to a boil, stirring constantly, until it is the consistency of pudding.

4. Add salt and allow mixture to cool.

5. Pour into airtight container and refrigerate.

■ Make Your Own Sticky Notes

NOTE: You can make your own sticky notes with ease. Repositionable glue sticks can be expensive, so look for sales. These sticky notes make great holiday gifts, Mother's Day gifts, or birthday presents.

Materials

repositionable glue sticks
scissors

multiple copies of colorful or interesting papers

Procedure

1. Apply some of the repositionable glue to the top of backs of each paper.

2. Attach papers together.

3. Trim edges if necessary.

6
Paints

Paint is made by combining a pigment (color) with a binder (viscous material). The binder makes the pigment adhere to the support (paper, wood, or other materials). Four types of pigment are nontoxic and are thus safe for children to use. Food coloring is transparent. However, it will not always wash out of clothes, and it can temporarily stain fingers. Watercolors are transparent and usually washable. Tempera and poster paints are more opaque. They are reasonably priced and easy to obtain. They often wash out of clothing. Acrylics are expensive, and they can sometimes ruin clothing. In most cases, you can interchange the pigments to meet your needs.

Finger paint paper can be bought from most craft catalogs. However, freezer paper and shelf paper are easier to obtain and work for the most part just as well. Different types of watercolor paper exist, but most of these paints are designed to work on ordinary white drawing paper.

■ Transparent Finger Paint
[Makes 3 1/4 cups]

NOTE: This shiny, transparent paint can be used on either wet or dry paper. It dries more quickly than other finger paints. Leftover paint lasts about a week if refrigerated. It can be used for painting windows. However, it is not as easy to remove from glass surfaces as are other paints.

Materials

1 tablespoon (1 envelope) unflavored gelatin	mixing bowl
3 cups water	mixing spoon
1/2 cup cornstarch	stove or heating element
1/4 cup dishwashing liquid	pot
food coloring	small containers with lids

Procedure

1. Combine gelatin and 1/2 cup water in mixing bowl. Let thicken.
2. Combine cornstarch and 2 1/2 cups cold water in pot. The cornstarch should dissolve. Simmer and stir until thick. Remove pot from heat.
3. Add gelatin mixture and then dishwashing liquid to pot.
4. Pour into small containers. Add food coloring to desired shades.
5. Paint is ready to use when it cools to room temperature.

■ Somewhat-Opaque Finger Paint
[Makes 2 1/2 cups]

NOTE: This finger paint flows smoothly, and it can be used on either wet or dry paper. The flour produces a smell that may bother some children. A bit of vanilla or lemon extract will mask the smell.

Materials

1/2 cup all-purpose flour	pot or microwave-safe
2 cups water	container
1 tablespoon glycerin	mixing spoon
vanilla or lemon extract (optional)	small containers with lids
poster paints or dry tempera paints	
stove, heating element, or micro-	
wave oven	

Procedure

1. In the pot, combine flour with enough water to make a paste.

2. Add rest of water and stir over low heat until mixture is thick. Or microwave mixture in microwave-safe container at high setting for 1 1/2 minutes, stir, and microwave again at high setting for another 1 1/2 minutes.

3. Add glycerin and flavoring.

4. Pour into small containers and let cool. Add poster paints or tempera paints to obtain desired shades.

■ Opaque Finger Paint
[Makes 1 cup]

NOTE: This paint is fun to feel and easy to use. It will layer over other colors.

Materials

1 cup liquid starch	mixing bowl
2 tablespoons cornstarch	mixing spoon
acrylic paints	small containers with lids

Procedure

1. Combine liquid starch and cornstarch in mixing bowl.

2. Pour into small containers and add acrylic paints until desired color is reached.

■ Smelly Finger Paint
[Makes 2 tablespoons of one color finger paint]

NOTE: These paints obviously have nice smells. They feel good on fingers; and the final product has a shiny, grainy finish. Substitute 1 tablespoon (1 envelope) unflavored gelatin, several drops of food coloring, and several drops of flavoring for the flavored, sugar-free gelatin if you want.

Materials

1 small package flavored, sugar-free gelatin
2 tablespoons hot water

small mixing bowl
mixing spoon

Procedure

1. Combine gelatin and water in the mixing bowl.
2. Let cool for about 15 minutes.
3. Use paints in one setting; these paints cannot be stored.
4. Works will dry usually overnight.

■ Classic Finger Paint
[Makes about 2 1/2 cups]

NOTE: This very old recipe is worth the effort.

Materials

1 cup all-purpose flour
3 tablespoons salt
1 1/2 cups cold water
1 1/4 cups hot water
food coloring

pot
stove or heating element
mixing spoon
small containers with lids

Procedure

1. Combine flour, salt, and cold water in pot.
2. Stirring constantly, heat mixture.
3. Slowly add the hot water.
4. Stirring constantly, bring mixture to a boil. Remove from the heat when it is thick.
5. Pour some of the mixture into each of several containers. Add food coloring until desired shade is reached.
6. Store in refrigerator.

■ Soap Flake Finger Paint
[Makes about 1 cup]

NOTE: This paint works well on dark paper.

Materials

1/4 cup soap flakes (not laundry detergent)
1/2 cup warm water
food coloring

egg beater
bowl
small containers with lids

Procedure

1. Pour water into bowl.
2. Slowly add a small amount of the soap flakes to the water. Beat.
3. Repeat the process until all the soap flakes have been added and concoction looks like whipped cream.
4. Divide mixture into several small containers and add food coloring until desired shade is reached.

■ Window Finger Paint
[Makes 1 1/4 cups]

NOTE: This finger paint is nice and thick. It produces an opaque picture on windows, and it is very easy to wash off.

Materials

1 cup baby shampoo
1/2 cup cornstarch
mixing bowl

mixing spoon
small containers
food coloring

Procedure

1. Combine baby shampoo and cornstarch in the mixing bowl.
2. Divide mixture into several small containers and add food coloring until desired shade is reached.

■ Cornstarch Finger Paint
[Makes about 2 1/2 cups]

NOTE: The glycerin gives it a nice flow.

Materials

1/2 cup cornstarch	pot
3/4 cup cold water	stove or heating element
2 cups hot water	mixing spoon
1 tablespoon glycerin	small containers with lids
food coloring or tempera pigments	

Procedure

1. Combine 1/4 cup cold water with cornstarch to make a paste.
2. Stir in hot water.
3. Heat at a low temperature, stirring constantly until mixture starts to boil.
4. Add 1/2 cup cold water and glycerin.
5. Pour into small containers and add food coloring or tempera pigments until desired colors are reached.

■ Laundry Starch Finger Paint
[Makes 2 cups]

NOTE: Laundry starch can be found in the detergents aisle of the grocery store.

Materials

1/4 cup cold water	pot
1/4 cup laundry starch	stove or heating element
1 1/2 cup hot water	spoon
2 tablespoons talcum powder	small containers with lids
1/4 cup soap powder	*Boric acid should not be eaten. If
1/2 teaspoon boric acid (for preservative)*	young children use this material, consider leaving the boric acid out.
food coloring or tempera pigments	

Procedure

1. Combine laundry starch and cold water in pot.
2. Add hot water and stir.
3. Over low heat cook until mixture is thick. Stir constantly.
4. Remove from heating element and add talcum powder, soap powder, and boric acid.
5. Pour into small containers and add food coloring or tempera pigments until desired colors are reached.

■ Edible Finger Paint
[Makes 4 cups]

NOTE: This paint actually dries if it is applied lightly to paper. Youngsters of all ages enjoy this. Consider adding a flavoring to the plain yogurt.

Materials

1 container plain yogurt	spoons
food colorings	wax paper
small dishes	

Procedure

1. Spoon some of the plain yogurt into the small dishes.
2. To each dish add food coloring until desired shades are reached.
3. Children can finger paint on the wax paper and lick their fingers as they go.

■ Versatile Paint
[Makes 4 small containers of paint]

NOTE: Versatile paint makes a great finger paint as well as a poster-type paint. It has only 2 ingredients and will keep for a long time. It provides a transparent, flat finish.

Materials

2 cups liquid starch	mixing spoon
1 cup each of 4 colors of dry tempera paint	4 small airtight containers with lids

Procedure

1. Pour liquid starch into each of the small containers.
2. Add dry paints, a different color to each container.
3. Mix ingredients.

■ Milk Paint
[Makes 1 1/2 cups]

NOTE: The early American colonists used milk paint to give wood a stained look. Milk was cheap and easy to obtain. This version produces pastel but vibrant shades. The finish is flat and transparent. Layering it over other colors is fun. This paint does not last long even if it is refrigerated.

Materials

3/4 cup powdered nonfat milk
1/2 cup water
powdered tempera paints
mixing bowl

mixing spoon
several small airtight containers with lids

Procedure

1. Combine powdered milk and water in mixing bowl.
2. Pour into small containers and add powdered paints until desired color is reached.

■ Casein Paint
[Makes 3/4 cup]

NOTE: The milk provides the casein part. This recipe was used by people centuries ago.

Materials

1/2 cup warm water
2 tablespoons borax
1/4 cup powdered nonfat milk

powdered tempera pigments
airtight container
spoon

Procedure

1. Combine borax and water in airtight container.
2. Add powdered nonfat milk and stir completely. Mixture should be thick.
3. Add powdered tempera paints until desired color is reached.

■ Glue Paint
[Makes 1 3/4 cups]

NOTE: Use this opaque paint to cover waxy or plastic surfaces. It has a flat finish.

Materials
1 cup white glue
acrylic paints
mixing spoon

several small airtight containers
with lids

Procedure
1. Pour glue into small containers.
2. Add acrylic paints to obtain desired colors.

■ Mucilage Paint
[Makes 1 cup]

NOTE: Mucilage is a type of glue that has been around for many, many years. Mucilage glue can be purchased at craft stores or through the Internet. This paint sticks nicely to cardboard.

Materials
1/2 cup mucilage
1/2 cup honey

jar with lid
powdered pigments

Procedure
1. Combine mucilage and honey in jar. Shake until well mixed.
2. Add powdered pigments until the desire color is reached.

■ Egg Tempera Paint
[Makes 1 1/4 cups]

NOTE: Egg tempera is a medium that has been used by painters from Michelangelo to Andrew Wyeth. It is opaque, but it can be thinned to any level of transparency by adding water. Although it is thick enough to be a good paint for windows, it is not easy to remove. This paint does not last long even if it is refrigerated.

Materials

1/2 cup beaten egg yolks (4 egg yolks)
powdered tempera paints
1/2 cup water

mixing spoon
several small airtight containers with lids

Procedure

1. Pour beaten egg yolks into small containers.

2. Add powdered tempera paints

3. Thin with water to desired consistency.

■ Thick Paint
[Makes 1 1/3 cups]

NOTE: This paint covers almost anything. For example, it will cover lettering on cardboard boxes.

Materials

1 cup papier-mâché paste (see p. 53)
2 tablespoons powdered tempera paint
water

1 tablespoon powdered soap (not laundry detergent)
mixing bowl
mixing spoon
airtight container

Procedure

1. Combine papier-mâché paste and powdered tempera paint.

2. Add enough water to make a thick paint.

3. Add powdered soap.

4. Store in airtight container.

■ Opaque Window Paint
[Makes about 1 1/2 cups]

NOTE: This paint is easy to make and easy to clean off windows when the time comes. It also dries completely. Conversely, transparent window paint (see next recipe) remains rather gooey.

Materials

1 cup liquid dishwashing soap	mixing bowl
1 cup cornstarch	mixing spoon
food coloring	airtight containers

Procedure

1. Combine dishwashing liquid and cornstarch in mixing bowl.

2. Spoon into containers.

3. Add food coloring. Remember that the dry paint will be lighter in color than the wet paint.

■ Transparent Window Paint

NOTE: Children love working with petroleum jelly. This paint does not run, and it can easily be removed. However, it does not dry to a hard finish.

Materials

petroleum jelly	plastic spoons
food coloring	paper cups

Procedure

1. Spoon a small amount of petroleum jelly into each paper cup.

2. Add drops of food coloring to each cup and blend. On a window, the paint will appear lighter than it does on the cup, so make the shade darker to compensate.

■ Black and White Opaque Window Paint

NOTE: The colors black and white are hard to achieve with the above formulas. Because black and white are often used, the following recipes solve the problem. If children are painting windows to look like stained glass, the opaque black becomes a great "lead" to separate other colors. These paints never really harden.

Materials

zinc oxide ointment (for white paint; available at pharmacies)

black tempera pigment (for black paint)

mixing spoon

airtight containers

Procedure

1. Use plain zinc oxide ointment for white paint.

2. Add black tempera pigment to zinc oxide ointment to make black paint.

3. Store paints in airtight containers.

■ Sidewalk and Spritz Paint

[Makes enough for 4 containers of 1/4 cup each]

NOTE: To make spritz paint, pour paint into a spray bottle. The next rain will wash away these paints.

Materials

1 cup cornstarch

1 cup water

food coloring

small containers with lids

mixing spoon

Procedure

1. Combine 1/4 cup water with 1/4 cup cornstarch into each of 4 containers.

2. Add food coloring or powdered tempera paints until desired shade is reached.

■ Gouache
[Makes 3/4 cup]

NOTE: A gouache is a heavy, opaque watercolor. This recipe requires dextrin, which is found in artificial sweeteners. These can be purchased at a grocery store.

Materials

2 cups dextrin
4 tablespoons water
1/2 cup honey
2 teaspoons glycerin

1/2 teaspoon boric acid
  (as preservative)*
powdered tempera pigments
jar
spoon

*Boric acid should not be eaten. If young children use this material, consider leaving the boric acid out.

Procedure

1. Combine dextrin and water in jar.

2. Stir in honey, glycerin, and boric acid. Thoroughly combine all ingredients.

3. Add powdered pigments until desired color is reached.

■ Puffy Paints
[Makes about 1 cup]

NOTE: These paints make great accents to projects. Colored sand and glitter can be used as substitutes for plain sand.

Materials

1/2 cup salt
1/2 cup all-purpose flour
1/2 cup water
powdered tempera paints
1 tablespoon sand

mixing bowl
spoon
paper or plastic cups
plastic squeeze bottles

Procedure

1. Combine salt and flour in mixing bowl.

2. Add water and sand.

3. Divide mixture among four or more cups.

4. Add powdered tempera paints until desired shades are reached.

5. Pour each color into a plastic squeeze bottle.

6. Squeeze paints onto paper or hard surfaces.

7. Let dry at least overnight and perhaps longer.

8. Refrigerate leftover paints, but bring back to room temperature before using.

■ Scratch-and-Sniff Paints
[Makes 1 tablespoon]

NOTE: Children could make their own valentines and add a scratch and sniff accent.

Materials

2 tablespoons unsweetened powdered
    drink mix
1 tablespoon warm water

small paper cups
mixing spoon

Procedure

1. Combine drink mix and water in cup.
2. Paint.
3. Allow surface to dry for 24 hours before scratching and sniffing.

■ Shaving Cream Paint and Sculpting Material
[Makes enough for 10 children]

NOTE: The finished sculptures are delicate and temporary. This project is fun, and children enjoy the smell.

Materials

1 can white shaving cream
food coloring

small paper cups
plastic spoon

Procedure

1. Squirt some shaving cream into each of the paper cups.
2. Carefully mix in several drops of food coloring.
3. Create sculptures and let dry.

■ Dish Soap Paint
[Makes 2 1/2 tablespoons of one color]

NOTE: This paint can last a long time. It is incredibly easy to make and fun to use.

Materials

2 tablespoons clear dish soap
2 tablespoons powdered tempera paint
1 teaspoon water

small mixing bowl
spoon
small containers with lids

Procedure

1. Mix soap, powdered paint, and water in small bowl.
2. Store in small containers.

■ Syrup Paint
[Makes 3 tablespoons of one color]

NOTE: This paint actually dries, and it is very glossy. Colors are very vibrant. It can also be used as a finger paint.

Materials

3 tablespoons light corn syrup
food coloring
small mixing bowl

spoon
small containers with lids

Procedure

1. Combine light corn syrup and at least 6 drops of food coloring in small mixing bowl.
2. Store in small containers.

■ Pan Paints 1 (Glycerin)
[Makes about 2 tablespoons]

NOTE: The end product is a dry, foamy transparent paint that can be used like watercolors.

Materials

1 tablespoon white vinegar	mixing bowl
3 tablespoons baking soda	mixing spoon
1/2 teaspoon glycerin	several small containers
powdered tempera pigments	

Procedure

1. Combine white vinegar and baking soda in mixing bowl.
2. After foaming stops, add glycerin.
3. Pour into containers and add powdered tempera paints to desired colors. The dry paint will be lighter than the wet paint.
4. Let harden and use as watercolors.

■ Pan Paints 2 (Corn Starch and Light Corn Syrup)
[Makes 1/3 cup base to be divided into several portions, one for each color]

NOTE: This mixture will foam at first. This recipe lets the children design their own colors.

Materials

3 tablespoons cornstarch	food coloring
3 tablespoons baking soda	mixing bowl
3 tablespoons white vinegar	mixing spoon
2 teaspoons light corn syrup	several containers with lids

Procedure

1. Combine cornstarch, baking soda, white vinegar, and light corn syrup in the mixing bowl.
2. Place a small amount of the mixture into each of several containers.
3. Add food coloring to achieve desired shade.
4. Use as is or wait until they dry into pan paints.

■ Laundry Starch Paint
[Makes about 1 1/2 cups]

NOTE: Laundry starch can be found in the detergents aisle of the grocery store. This paint is easy to use.

Materials

1 cup laundry starch
1 cup soap powder (not laundry detergent)
3 tablespoons powdered tempera pigments

1 cup water
mixing bowl
mixing spoon
airtight container

Procedure

1. Combine laundry starch, soap powder, and powdered tempera paint in mixing bowl.
2. Add water and mix until smooth.
3. Store in airtight container.

■ Detergent Paint
[Makes about 3/4 cup]

NOTE: This paint has a grainy texture at first until the detergent fully dissolves.

Materials

1 cup powdered detergent
6 tablespoons liquid tempera paint
1/4 cup water

mixing bowl
mixing spoon
airtight container

Procedure

1. Combine powdered detergent and liquid tempera in the mixing bowl.
2. Add water until mixture is similar to pudding.
3. Pour into airtight container.

7

Face and Body Paints

Commercial face painting kits are expensive and limited in colors. Face paints can be made with readily available ingredients and a little imagination. Remember to test these face paints on a small patch of skin before making major designs. Some children could have a reaction to the materials.

■ Cornstarch Face Paint
[Makes 3 tablespoons]

NOTE: Keep food coloring away from clothing.

Materials

2 tablespoons shortening	mixing spoon
1 tablespoon cornstarch	cotton swabs, small brushes, or
3–6 drops food coloring	make-up sponges
small container	

Procedure

1. Combine shortening and cornstarch in small container.
2. Add food coloring until desired shade is reached.
3. Apply with cotton swabs, small brushes, or make-up sponges.
4. Remove with soap and water.

■ Body Paint
[Makes 3 tablespoons]

NOTE: Cold cream can be purchased at a drugstore or cosmetics supply store. It is very thick, and it easily combines with other ingredients. Keep food coloring away from clothing.

Materials

1 tablespoon cold cream	small container
1 tablespoon cornstarch	mixing spoon
1 tablespoon water	cotton swabs, small brushes, or
3–6 drops food coloring	make-up sponges

Procedure

1. Combine cold cream and cornstarch in small container.
2. Stir in water until mixture is creamy.
3. Add food coloring until desired shade is reached.
4. Apply with cotton swabs, small brushes, or make-up sponges.
5. Remove with soap and water.

■ Corn Syrup Face Paint

[Makes 2 tablespoons]

NOTE: This face paint can be applied easily to the face. Keep food coloring away from clothing.

Materials

1 tablespoon cold cream
1 tablespoon corn syrup
3–6 drops food coloring
small container

mixing spoon
cotton swabs, small brushes, or
   make-up sponges

Procedure

1. Combine cold cream and corn syrup in small container.
2. Add food coloring until desired shade is reached.
3. Apply with cotton swabs, small brushes, or make-up sponges.
4. Remove with soap and water.

■ Baby Shampoo Face Paint

[Makes 3 tablespoons]

NOTE: The tempera pigments could stain clothing.

Materials

3 tablespoons baby shampoo
powdered tempera paint
small mixing bowl

mixing spoon
cotton swabs, small brushes, or
   make-up sponges

Procedure

1. Pour baby shampoo into small mixing bowl.
2. Add powdered tempera paint until desired shade is reached.
3. Apply with cotton swabs, small brushes, or make-up sponges.
4. Remove with soap and water.

■ Baby Lotion Face Paint
[Makes 1/4 cup]

NOTE: This formula produces a smooth consistency.

Materials

1/4 cup baby lotion
powdered tempera paint
2 teaspoons dish soap
small mixing bowl

mixing spoon
cotton swabs, small brushes, or
   make-up sponges

Procedure

1. Pour baby lotion into small mixing bowl.
2. Add powdered tempera paint until desired shade is reached.
3. Stir in dish soap.
4. Apply with cotton swabs, small brushes, or make-up sponges.
5. Remove with water.

■ Three-Dimensional Makeup
[Makes 1 tablespoon]

NOTE: This face paint is great for Halloween, because it makes great bumps and wrinkles. This face paint must be used immediately, and any leftover paint should be thrown away and not poured down a sink.

Materials

1 tablespoon (1 envelope) unflavored
   gelatin
1 1/2 tablespoons hot water
food coloring and/or cocoa powder

old mixing bowl
fork
cotton swab or fine paintbrush

Procedure

1. In small mixing bowl, combine unflavored gelatin and hot water. Stir with fork until gelatin dissolves.
2. Let gelatin sit until mixture thickens.
3. Add food coloring and/or cocoa powder.
4. Apply makeup to face or body with more cotton swabs.

8

Natural Dyes

People have been dyeing textiles, paper, and other materials with natural stains for over 5,000 years. Three steps are involved in the natural dye process. First, the dye bath must be prepared. During this step, berries, leaves, stems, or roots are simmered or boiled in water. The color-bearing material is often crushed and then strained from the water, leaving the dye bath.

Second, the textile, paper, or other material to be dyed must be prepared. Most materials, such as clay, papier-mâché, or handmade paper, need little preparation. However, fibers from fabrics must be treated to keep the dye from fading, washing out, or flaking off. Fibers are simmered in a chemical solution called a mordant. The mordant affects the final color as well. Some mordants are poisonous, and so they must be handled with care. Children should not work with mordants. Therefore, none of these recipes uses mordants. Fabrics dyed using the following recipes may not be colorfast.

In the third step, the material to be dyed is steeped in the dye bath. Sometimes the material is simmered or boiled in the dye bath. Usually the longer the material is in the dye bath, the darker the shade will be. Remember again that the fabrics may not be colorfast, so wash with care.

All dye baths should be prepared in enamel or stainless steel pots. Wooden spoons should be used to stir concoctions.

The following formulas are arranged according to ROY G. BIV and the brown family.

■ Cranberry Dye (Bright Red Color)
[Makes 1 quart]

NOTE: Cranberries are easy to find around Thanksgiving.

Materials

2 cups cranberries
1 quart water
large enamel or stainless steel pot
wooden mixing spoon

stove or heating element
sieve
container to collect dye bath
material to dye

Procedure

1. Combine cranberries and water in pot. Simmer for 15 minutes.

2. Crush berries and simmer for 15 more minutes.

3. Strain mixture and discard berries.

4. Add material to dye. Let soak until desired color has been reached.

■ Beet Dye (Red Color)
[Makes 1 1/2 cups]

NOTE: Consider using beet juice if you can find it.

Materials

2 15-ounce cans cooked beets
sieve

container to collect dye bath
material to dye

Procedure

1. Open cans of beets and strain.
2. Use juices as the dye bath.
3. Add material to dye. Let soak until desired color has been reached.

■ Carrot–Paprika Dye (Orange Color)
[Makes 2 cups]

NOTE: Paprika is a mix of different types of ground, dried peppers. Different types of paprika will provide different shades of red.

Materials

1 cup grated carrots
4 tablespoons paprika
2 cups water
pot
wooden spoon

stove or heating element
sieve
container to collect dye bath
material to dye

Procedure

1. Cook carrots in water for about 15 minutes.
2. Add paprika and thoroughly combine.
3. Pour material into sieve. Use back of spoon to press out any remaining dye.
4. Strain mixture and discard cooked carrots.
5. Add material to dye. Let soak until desired color has been reached.

■ Marigold Dye (Yellow Color)
[Makes 1 quart]

NOTE: Marigolds are plentiful in late summer. The marigold petals must be soaked overnight.

Materials

2 cups marigold petals (collected at height of bloom)
1 quart water
large enamel or stainless steel pot
wooden mixing spoon

stove or heating element
sieve
container to collect dye bath
material to dye

Procedure

1. Combine marigold petals and water in pot. Soak overnight. Simmer for 15 minutes.
2. Crush petals and simmer for another 15 minutes.
3. Strain mixture and discard marigold petals.
4. Add material to dye. Let soak until desired color has been reached.

■ Mustard Dye (Yellow Color)
[Makes 1 quart]

NOTE: This dye is mustard-yellow in color.

Materials

1/2 cup prepared mustard
1 quart water
wooden mixing spoon

container to collect dye bath
material to dye

Procedure

1. Combine mustard and water in container.
2. Add material to dye. Let soak until desired color has been reached.

■ Onion Skin Dye (Yellow Color)
[Makes 1 quart]

NOTE: This dye is yellow to brown in color, depending on the onions.

Materials

1/2 gallon loosely packed, dry skins
    from yellow onions
1 quart water
large enamel or stainless steel pot
wooden mixing spoon

stove or heating element
sieve
container to collect dye bath
material to dye

Procedure

1. Tear onion skins into small pieces.
2. Boil onion skins in 1 quart water for 30 minutes.
3. Press skins to get out all the color.
4. Strain mixture and discard skins.
5. Let solution cool.
6. Add material to dye. Let soak until desired color has been reached.

■ Dill Seed Dye (Gold Color)
[Makes 2 cups]

NOTE: Dill and dill seeds are used in many, many cultures as herbs and herbal medicines.

Materials

1/4 cup dill seeds
2 cups water
large enamel or stainless steel pot
wooden mixing spoon

stove or heating element
sieve
container to collect dye bath
material to dye

Procedure

1. Boil dill seeds in 2 cups water for 30 minutes.
3. Press seeds to get out all the color.
4. Strain mixture and discard seeds.
5. Let solution cool.
6. Add material to dye. Let soak until desired color has been reached.

■ Spinach Dye (Green Color)
[Makes 1 quart]

NOTE: Bigger, tougher spinach leaves are best for this dye.

Materials

2 cups chopped spinach

1 quart water

large enamel or stainless steel pot

wooden mixing spoon

stove or heating element

sieve

container to collect dye bath

material to dye

Procedure

1. Combine spinach and water in pot. Simmer for 15 minutes.

2. Crush spinach and simmer for 15 more minutes.

3. Press spinach to get out all the color.

4. Strain mixture and discard spinach.

5. Add material to dye. Let soak until desired color has been reached.

■ Purple Cabbage Dye (Blue Color)
[Makes 1 quart]

NOTE: Some people dislike the smell of cooked cabbage.

Materials

2 cups chopped purple cabbage

1 quart water

large enamel or stainless steel pot

wooden mixing spoon

stove or heating element

sieve

container to collect dye bath

material to dye

Procedure

1. Combine cabbage and water in pot. Simmer for 15 minutes.

2. Crush cabbage and simmer for 15 more minutes.

3. Press cabbage to get out all the color.

4. Strain mixture and discard cabbage.

5. Add material to dye. Let soak until desired color has been reached.

■ Blueberry Dye (Bluish Purple Color)
[Makes 1 quart]

NOTE: Blueberries are cheapest in early summer.

Materials

2 cups blueberries
1 quart water
large enamel or stainless steel pot
wooden mixing spoon

stove or heating element
sieve
container to collect dye bath
material to dye

Procedure

1. Combine blueberries and water in pot. Simmer for 15 minutes.
2. Crush blueberries and simmer for 15 more minutes.
3. Press berries to get out all the color.
4. Strain mixture and discard berries.
5. Add material to dye. Let soak until desired color has been reached.

■ Grape Juice Dye (Purple Color)
[Makes 1 quart]

NOTE: This dye does not have to be cooked.

Materials

1 quart grape juice
container to collect dye

material to dye

Procedure

1. Pour juice into container.
2. Add material to the dye. Let soak until desired color has been reached.

■ Tea Dye (Tan Color)
[Makes 1 quart]

NOTE: Use black teas and not green teas for this dye.

Materials

5 tea bags
1 quart water
large enamel or stainless steel pot
wooden mixing spoon

stove or heating element
sieve
container to collect dye bath
material to dye

Procedure

1. Combine tea bags and water in pot. Boil for 15 minutes.
2. Strain mixture and discard tea bags.
3. Add material to dye. Let soak until desired color has been reached.

■ Walnut Shell Dye (Brown Color)
[Makes 1 quart]

NOTE: The walnut shells must be soaked overnight.

Materials

2 cups walnut shells
1 quart water
large enamel or stainless steel pot
wooden mixing spoon

stove or heating element
sieve
container to collect dye bath
material to dye

Procedure

1. Combine walnut shells and water in pot. Soak overnight. Boil for 1 hour.
2. Strain mixture and discard walnut shells.
3. Add material to dye. Let soak until desired color has been reached.

■ Coffee Dye (Brown Color)
[Makes 1 quart]

NOTE: Used coffee grounds work nicely for this dye.

Materials

1 cup coffee grounds
1 quart water
large enamel or stainless steel pot
wooden mixing spoon

stove or heating element
sieve
container to collect dye bath
material to dye

Procedure

1. Combine coffee grounds and water in pot. Simmer for 15 minutes.

2. Strain mixture and discard coffee grounds.

3. Add material to dye. Let soak until desired color has been reached.

9

For the Birds

Children really enjoy watching birds. They love to record types and numbers of birds, and looking for bird's nests can become a passion. Children can learn more about birds by providing bird food and making bird feeders.

Birds can be both finicky and fickle when it comes to feeding time. For example, blue jays prefer sunflower seeds, cracked corn, and shelled peanuts. However, they will also eat doughnuts and crackers. Woodpeckers choose suet and bacon drippings, but they will also consume shelled peanuts if that is all there is.

The most popular kinds of bird food are sunflower seeds, cracked corn, suet, and niger. Birds that prefer sunflower seeds seem to like the gray-striped variety best. Suet, a white, solid animal fat, can be obtained from the butcher. Place a piece of suet in a mesh bag, such as those grapes come in, or in a suet cage and hang it from a tree branch. Niger, also called thistle, is far more expensive than other kids of seeds. Therefore, some people prefer to buy seed blends. Birds also like peanut butter and small pieces of fruits and nuts. Popped popcorn and other seeds, such as those from pumpkins, melons, peppers, and so on, attract birds.

Many birds need grit to help grind food in their digestive tracts. One easy way to provide grit is to crush egg shells. Put clean egg shells in a plastic bag and seal. Then use a rolling pin to roll over and over the shells. It will not take long to break them up. You may also use fire ashes, sand, or poultry grit.

Birds also like water, even in winter. A hose dripping into a bucket is enough to attract many winged friends.

Finally, coarse salt will bring birds. Place the grit and salt a slight distance away from the food so that birds can choose from a smorgasbord.

Once you start to feed birds, stick with the plan. Birds become accustomed to the food supply. They particularly need food in late winter and early spring.

Bird Food

■ Homemade Bird Seed Mixture
[Makes 4 cups]

NOTE: This bird seed can be cheaper than commercial brands. It can also be tailored to specific needs. No expensive niger is in this recipe. Add 3 to 5 crushed egg shells or 1/2 cup sand for grit.

Materials

1 cup black-oil sunflower seeds
1 1/4 cup cups striped sunflower
   seeds
1/2 cup sunflower hearts
1/2 cup millet

1/2 cup dried corn
1/4 cup safflower seeds
mixing bowl
mixing spoon
airtight container

Procedure

1. Combine all ingredients in mixing bowl.

2. Distribute to various feeders.

3. Store seed in airtight container.

■ Bird Food 1 (Muffins for Birds)
[Makes 5 1/4 cups]

NOTE: Almost every type of bird enjoys this food. The recipe is easy, and children can do all the work.

Materials

1 pound melted suet or shortening
1 cup chunky peanut butter
1 cup rolled oats
1 cup cornmeal
1 cup niger

1 cup sunflower seeds
mixing bowl
mixing spoon
24 paper muffin liners
muffin pans

Procedure

1. Stir all ingredients together.

2. Pour into paper-lined muffin pans. Let mixture harden.

3. Peel away paper liners before serving to birds.

■ Bird Food 2 (Food Balls for Birds)
[Makes 8 1/2 cups]

NOTE: Because this mixture does not have to dry, children can make the bird food balls and take them outside immediately.

Materials

2 cups bread crumbs
1/2 pound melted suet
3 chopped apples, including skin
    and seeds
1/2 cup all-purpose flour
1/4 cup cornmeal

1/2 cup raisins
1/2 cup chopped nuts
1 cup peanut butter
1 cup wild bird seed
mixing bowl
mixing spoon

Procedure

1. Combine all ingredients in mixing bowl.

2. Shape into balls and put them out for birds.

■ Bird Food 3 (Bird Food Free from Bird Seed)
[Makes 6 cups]

NOTE: This recipe uses only ingredients found in the kitchen. It does not include any bird seed.

Materials

1 cup cornmeal
1 cup rolled oats
1 cup flour
1 cup wheat germ
1 cup raisins
1/2 cup shortening

1 cup powdered nonfat milk
mixing bowl
mixing spoon
baking pan
nonstick cooking spray
oven

Procedure

1. Combine all ingredients in mixing bowl.

2. Spray pan with nonstick cooking spray.

3. Pour in the batter. Bake at 350°F for 1 hour.

4. Let cool. Break into walnut-sized pieces.

■ Bird Food 4 (Popsicles for Birds)
[Makes 1]

NOTE: Birds need water even in the winter. They can get some moisture from the melting water. Birds also like these popsicles on hot days!

Materials

birdseed
cranberries
large paper cup

scrap of string or yarn
water
freezer

Procedure

1. Pour some birdseed and cranberries into the large paper cup.

2. Fill with water.

3. Place the 2 ends of the string into the water to form a handle.

4. Place in the freezer for about a day.

5. When frozen, take out and hang on a branch.

6. Watch the birds come for their frozen treats.

■ Bird Food 5 (Treats for Birds)

[Makes 1]

NOTE: Save the bread scraps and let them dry. Crush the scraps into bread crumbs for other projects.

Materials

slice of bread   large cookie cutter
peanut butter   drinking straw
birdseed   string

Procedure

1. Apply the cookie cutter to the slice of bread and remove the shape.
2. Spread peanut butter on the design.
3. Sprinkle with birdseed.
4. Poke a hole at the top with the drinking straw.
5. Let the shapes dry overnight.
6. Put a piece of string through the hole and tie to a branch.
7. Stand back and watch the birds enjoy their treat.

■ Bird Food 6 (Energy Cakes for Birds)

[Makes 5]

NOTE: Birds need high-calorie foods, especially in winter. This mixture gives them the calories to keep up their energy.

Materials

2 cups melted shortening   old mixing spoon
1 cup peanut butter   5 empty, clean yogurt containers
2 cups corn meal   knife
old mixing bowl   string

Procedure

1. Combine the melted shortening and peanut butter in the old mixing bowl.
2. Add the corn meal slowly and thoroughly combine with the shortening and peanut butter.
3. Use the knife to punch 2 small holes near the top of each yogurt container.
4. Thread string through the holes and tie knots in the strings.
5. Spoon some of the mixture into each of the yogurt containers.
6. Allow the mixture to cool.
7. Tie the filled containers to tree branches or bird feeders.

■ Bird Food 7 (S'Mores for Birds)
[Makes 6]

NOTE: Most birds love graham crackers.

Materials
6 graham crackers

1/2 cup peanut butter

1/2 cup raisins

knife

Procedure
1. Spread the peanut butter on the graham crackers.
2. Sprinkle raisins on top of the peanut butter.
3. Place the s'mores in bird feeders.

■ Bird Food 8 (Fruit Kabobs for Birds)
[Makes 6]

NOTE: Each fruit kabob also provides a perch for the birds so that they can rest and eat.

Materials
about 2 pounds of fruit pieces (such as apples, peaches, pears, or plums)

6 pieces of string each about 30 inches long

knife

6 very small disposable pie plates

Procedure
1. Use the knife to punch a small hole through the center of the bottom of each pie plate.
2. Thread the string through the bottom and make a knot. The remaining string rests in the pie plate.
3. Use the knife to make small holes in the pieces of fruit.
4. Thread the string through the fruit.
5. Use the leftover string to tie the kabobs onto tree branches.

■ Hummingbird Food
[Makes 4 cups]

NOTE: To first attract hummingbirds, make the ratio of sugar to water 1 to 3. After the hummingbirds become regular customers, change the ratio to 1 to 4. Wash the hummingbird feeder every week. Hummingbirds are attracted to the color red. Make sure some portion of the hummingbird feeder is bright red.

Materials

1 cup sugar   pot
4 cups water   hummingbird feeder
stove or heating element

Procedure

1. Combine sugar and water in pot and boil for 2 minutes.

2. Let solution cool and place into hummingbird feeder.

Bird Feeders

Birds are particular about foods, and they are selective about where they eat as well. All bird feeders should be near protective bushes or trees.

Many birds (e.g., mourning doves and juncos) feed on the ground. Simply clear away fallen leaves or snow from a flat area and scatter some cracked corn.

Another way to feed birds is to make feeder trays. Place old cafeteria trays on bricks or stumps. Spread some cracked corn, bread crumbs, or any of the above bird foods. Occasionally wash the trays.

Suet attracts many insect-eating birds such as woodpeckers. Place suet pieces in mesh bags such as those that contain produce. Garden supply stores also sell suet cages. Hang the bags or cages from tree branches.

The following bird feeders hang from poles or tree branches or are tied to a tree trunk. They use recycled materials and are easy to make.

■ Pinecone Bird Feeder

NOTE: This recipe is messy but fun. Children can set up an assembly line and work on these pinecone feeders as a cooperative project.

Materials

newspaper	table knife
pinecones	wax paper
peanut butter	string
assorted toppings, such as raisins, sunflower seeds, cracked corn, or cornmeal	

Procedure

1. Cover work area with pieces of newspaper.
2. Spread peanut butter on pinecones with knife. Make sure to get peanut butter down inside the grooves.
3. Place pieces of wax paper on newspaper. Pour toppings on wax paper.
3. Roll peanut butter-covered pinecones in toppings.
4. Tie string at stem end of pinecones and hang pinecones from tree branches.

■ Birds' Holiday Tree

NOTE: The tradition of a birds' holiday tree goes back more than 400 years. This project is a good way for children to help the environment and have fun at the same time.

Materials

popped popcorn	raisins
raw cranberries	sunflower seeds
cold cereal with holes	niger
(e.g., Cheerios®)	small pieces of bread or crackers
needle and thread	mixing bowl
oranges, cut in half	mixing spoon
grapefruit, cut in half	string

Procedure

1. Pick out a good tree within view of room.

2. Make garlands from the thread, popcorn, cranberries, and cereal. Hang garlands from tree.

3. Scoop out pulp from oranges and grapefruit. Mix pulp with raisins, sunflower seeds, niger, and small pieces of bread or crackers.

4. Spoon a bit of the pulp mixture back into each orange or grapefruit half. Attach strings to shell and hang shells from tree.

■ Bird Feeder 1 (Recycled Soda Bottle)
[Makes 1 bird feeder]

NOTE: This bird feeder is easy to refill. Children can see how much bird food is consumed in a day.

Materials

empty, clean 1-liter plastic soda bottle
plastic plate
heavy string
small, sharp knife

hot glue gun and glue
hammer
large nail
sunflower seeds to fill bottle

Procedure

1. Take off reinforcement from bottom of soda bottle. Punch holes along bottom of bottle with knife. The holes should be big enough for the seeds to fall through.
2. Hot glue bottom of bottle to plastic plate to form perch for birds.
3. Punch hole in top of cap with hammer and nail. Place some string through the hole. Tie a knot in the string inside the cap so that it will not fall through.
4. Fill bottle with sunflower seeds. Screw on cap.
5. Tie free end of string around tree branch.
6. Wash feeder occasionally.

■ Bird Feeder 2 (Recycled Milk Container)
[Makes 1 bird feeder]

NOTE: By recycling the plastic bottle, children are helping to preserve their environment.

Materials

plastic bottle with a distinct handle,
   such as a gallon milk container or
   a 1/2-gallon ammonia container

craft knife
wire
bird food

Procedure

1. Thoroughly wash and dry plastic container.
2. On the side away from handle, cut away a large square of plastic. The bottom edge of the cut will serve as a perch while birds eat.
3. Using wire, tie bottle to a tree trunk. Loop wire over a low branch before twisting wire ends together.
4. Fill bottle with bird food.

■ Bird Feeder 3 (Simple String Feeder)
[Makes 1 bird feeder]

NOTE: Any food that has a hole in the middle is possible food for this feeder. Orioles particularly like orange slices. Rain is the enemy of this bird feeder.

Materials

1 piece of string about a yard long orange slices

1 bagel sliced thinly

Procedure

1. Thread string through the centers of the orange slices and bagel slices.

2. Tie the string around a branch of a tree.

■ Bird Feeder 4 (Plastic Plant Hanger)
[Makes 1 bird feeder]

NOTE: A plastic plant hanger serves as the basis for the bird feeder. Birds can perch on the lip of the drainage tray and nibble at the food that falls through the drainage holed. This feeder is easy to make and easy to refill.

Materials

1 plastic plant hanger with attached drainage tray
1 cardboard disk with diameter slightly smaller than that of the plant hanger

aluminum foil to cover cardboard

Procedure

1. Cover cardboard disk with aluminum foil. This will serve as the cover to the bird feeder.

2. Place bird food in the plant hanger. Some of the food will fall through the drainage holes into the drainage tray.

3. Cover the opening of the plant hanger with the cardboard disk.

4. Hang the planter from a pole or tree branch.

■ Springtime Nest Building Material

NOTE: Place the materials outside a window so that the children can see the birds in action. Children may want to save their dryer lint over winter to have enough for the spring.

Materials

scraps of yarn or string about
5 inches long

lint from clothes dryer
small twigs and branches

Procedure

1. Loosely drape scraps of string or yarn to branches of trees.
2. Place the dryer lint, twigs, and branches where birds can reach it.
3. The birds will use this material to build their nests.

10
For the Bugs

Did you know that over 900,000 species of insects exist? That number alone fascinates children, let alone the bugs themselves. The following activities help children learn about insects.

■ Butterfly Cage (Temporary Home)
[Makes 1]

NOTE: This is really an activity for 2 people. It is difficult to keep embroidery hoops in the right places. Netting, which comes in quite a few lovely colors, can be purchased at a fabric store.

Materials

1 yard netting

2 embroidery hoops

8 pieces of yarn, each 12 inches long

1 piece of yarn about 2 feet long

Procedure

1. Spread netting out on a flat surface.
2. Place 1 embroidery hoop in middle of netting.
3. Gather netting around hoop.
4. Place the other hoop about 6 inches above the first hoop.
5. Tie the second hoop to the first hoop with 4 pieces of yarn.
6. Continue to draw up the netting until a bag has been formed. Attach top hoop to netting with rest of yarn.
7. Tie the netting together with the long piece of yarn. Then suspend the cage from a wire.

■ Critter Cage (Temporary Home)
[Makes 1]

NOTE: Netting could be substituted for panty hose.

Materials

1 quart milk carton, empty and clean
old panty hose

rubber band
scissors

Procedure

1. Cut two openings in opposite sides of the milk carton.
2. Add whatever materials your critter might need (water, leaves, etc.).
3. Cut off a leg of the panty hose and put the milk carton inside the leg.
4. Add your critter.
5. Pull the panty hose leg completely over the milk carton and fasten shut with the rubber band.
6. Trim off any excess panty hose.

■ Moth Attractor Goo
[Makes 1 batch]

NOTE: Moths are for the most part night insects. The attractor goo can be prepared during the day, but a night visit is essential to seeing the moths.

Materials

2 cups orange juice that has been at
    room temperature for about 2 days
4 over-ripe bananas
1/2 cup honey or corn syrup
mixing bowl

mixing spoon
plastic wrap
old paintbrush
flashlight

Procedure

1. Place peeled bananas in the bowl and mash with the back of the mixing spoon.
2. Add orange juice and combine.
3. Add honey or corn syrup and combine.
4. Cover bowl with plastic wrap and allow it to sit outside in the sun for several hours.
5. Take the mixture and old paintbrush to a clearing where several trees are next to open space.
6. Paint several tree trunks with the mixture.
7. After dark, return to the area and see if the moths like the mixture.

■ Butterfly Nectar

[Makes 1 quart, enough for 1 feeder for about 3 weeks]

NOTE: Butterflies seek out nectar from flowers. This nectar satisfies their taste buds as well. Butterflies are attracted to bright colors, especially orange and purple. Therefore, color the butterfly feeder (nectar container) bright orange and purple. Clean the container often so that the nectar does not spoil.

Materials

1 quart water
1 cup sugar
old pot
stove or heating element
mixing spoon
small, empty, clean cottage
   cheese container

sponge cut to the size of the cottage
   cheese container (hopefully
   painted orange or purple)
airtight container

Procedure

1. Pour water into pot and heat until it boils.
2. Turn down heat and add sugar. Stir and cook until all the sugar is dissolved.
3. Remove pot from the heat and allow solution to cool.
4. Place sponge in empty cottage cheese container.
5. Pour in enough nectar to more than cover sponge.
6. Place the container on a rock or ledge in a sunny part of the garden. Wait for your guests!
7. Refrigerate remaining nectar in airtight container.
8. Replace nectar every 3–4 days.

■ Food for All Kinds of Bugs

[Makes 1 cup, enough for about 3 days]

NOTE: This food attracts bees and wasps as well as butterflies and ants. Therefore, children must watch out for stinging insects.

Materials

1 cup over-ripe fruit, cut into pieces   1 small, disposable pie pan

Procedure

1. Place the over-ripe fruit in the disposable pie pan.
2. Put the pie pan on a rock or large piece of wood. Give your insect guests time to find it.

11
Growing Plants

Children love growing plants, and they can learn so much in the process! What conditions are necessary for growth? What kinds of plants exist? How do plants reproduce? What kinds of plants do we eat? The list of possibilities goes on and on. Here are some easy ways to make plants an important part of learning.

■ Sweet Potato Plant
[Makes 1 plant]

NOTE: The sweet potato vine can become quite long. One year the vine went around the classroom! Children could incorporate some math into the activity by charting its weekly growth.

Materials

1 sweet potato
4 round or square sturdy toothpicks

1 tall, clear plastic glass
water

Procedure

1. Hold sweet potato so that the pointed ends are vertical. Poke 4 toothpicks into the sweet potato on all four sides of the middle.
2. Fill plastic glass with water.
3. Place one pointed end of the sweet potato into the water so that the 4 toothpicks rest on the edge of the glass. The toothpicks keep the top half out of the water. The top half of the sweet potato should remain dry, and the bottom half should be submerged in water.
4. Place the glass in a sunny spot and wait. Replace water as needed.
5. In about a week roots will grow from the bottom, and leaves will appear from the top.
6. When the sweet potato develops vines, transplant the plant to a pot filled with soil.

■ Potato Plant
[Makes about 5 plants, depending on the potato]

NOTE: Children could find out what part of the plant the actual potato is. Children could also research types of potatoes.

Materials

1 potato with plenty of eyes   knife
5 hot beverage cups   old aluminum tray
potting soil

Procedure

1. Cut potato into pieces so that each piece has at least 1 eye.
2. Use knife to cut a small hole in the bottom of each cup so that extra water can drain out.
3. Pour enough potting soil into each cup to form a layer about an inch thick.
4. Place a piece of the potato in each cup and fill in with more potting soil.
5. Place the cups on the old aluminum tray and water. Make sure the soon-to-be plants get plenty of sun and enough but not too much water.
6. The new plants should appear within a few days.

■ Carrot Plant
[Makes 1 plant]

NOTE: The green leaves are the key to this project. Are we cloning the carrot?

Materials

1 carrot with green leaves and stems   water
1 old aluminum pie pan

Procedure

1. Cut off the top of carrot so that about an inch of the carrot is still attached to the stem and leaves.
2. Place top in a pan of water. Keep adding water as it evaporates.
3. The new plant should begin to grow in about a week.

■ Pineapple Plant
[Makes 1 plant]

NOTE: Is the pineapple part that we eat a fruit? Is it a stem? Is it a sort of leaf?

Materials

1 pineapple with green leaves
   and stem
1 old aluminum pie pan

soil
water

Procedure

1. Cut off the top of the pineapple so that about an inch of the fruit is still attached to the stem and leaves.
2. Pour about an inch of soil into the aluminum pie pan.
3. Place the top of the pineapple into the pan of soil. Moisten and keep adding water as it evaporates.
4. The new plant should begin to grow in about a week.

■ Avocado Plant
[Makes 1 plant]

NOTE: Children could investigate the nutritional value of the avocado. They could also make some guacamole from the flesh of the avocado.

Materials

1 avocado pit
4 round or square sturdy toothpicks

1 tall, clear plastic glass
water

Procedure

1. Poke the 4 toothpicks into the avocado pit on all four sides of the middle.
2. Fill plastic glass with water.
3. Place avocado pit into the water so that the 4 toothpicks rest on the edge of the glass. The toothpicks keep the top half out of the water. The top half of the pit should remain dry, and the bottom half should be submerged in water.
4. Place the glass in a sunny spot and wait. Replace water as needed.
5. In about a week roots will grow from the bottom, and leaves will appear from the top.
6. When the avocado develops roots, transplant the plant to a pot filled with soil.
7. When the plant is about 8 inches tall, cut off the top half of the leaves. This cut promotes development of new branches.

■ Ginger Plant
[Makes 1 plant]

NOTE: Ginger root can be purchased from the produce section of a grocery store. Fresh ginger is used quite a bit in Asian cooking and medicine.

Materials

1 piece of ginger root about 3 inches long
1 small pot and saucer

potting soil
trowel

Procedure

1. Place about 1 inch of potting soil in the small pot.
2. Place 1 end of the ginger root into the soil and surround the root with more potting soil.
3. Make sure the root emerges from the top of the soil.
4. Water and place in direct sunlight. A plant should appear within 2 weeks.

■ Plants from Seeds
[Makes enough for 1 experiment]

NOTE: Children can watch the development of the root system in this activity.

Materials

1 large clear plastic cup
seeds (lima bean or radish seeds
   are good)

paper towels
water

Procedure

1. Line the plastic cup with paper towels.
2. Pour about an inch of water into the cup.
3. Place the seeds between the inside of the plastic cup and the paper towels.
4. Place the plastic cup in a sunny place. Make sure to keep water in the bottom of the cup.
5. In about a week the seeds should germinate. Children can observe the process of germination. They can see the beginning root system, the early stem, and baby leaves.
6. Transplant the seeds to pots with soil or plant outside.

■ Grass Heads
[Makes 1 head]

NOTE: Children can chart the growth of the grass. They can pull out some of the roots and examine them under a hand lens.

Materials

1 hot beverage cup
colored markers
potting soil

grass seeds
water

Procedure

1. The cup will be the head, so a face can be drawn on the cup with markers.
2. Fill the cup about 2/3 full with potting soil.
3. Scatter grass seeds over the surface of the potting soil.
4. Cover the seeds with a small amount of soil.
5. Moisten the soil with water and set the beverage cup in the sun.
6. After several days grass will appear as "hair."
7. Children can trim the grass "hair" with scissors.

■ Chia Critters
[Makes 1]

NOTE: Chia seeds, when milled, can be added to salads or cereals. Many people extol the nutritious virtues of chia seeds. A chia critter is much cheaper than the commercial version.

Materials

2 tablespoons chia seeds
disposable beverage cup
1 old knee-high panty hose
about 1 cup potting soil
rubber band

small pot
water
craft items such as pipe cleaners or
    googly eyes

Procedure

1. Pour seeds in disposable beverage cup. Add enough water to more than cover the seeds. Let seeds soak overnight. The seeds will germinate faster if this step is included.

2. The next day discard water. Seeds will be sticky.

3. Pour chia seeds into toe of panty hose. Eventually the toe will become the top of the critter's head.

4. Pour in potting soil and close end with rubber band. Extra panty hose can be trimmed off.

5. Turn critter over. The hose toe is now the top of the head. Children can add features such as googly eyes or pipe cleaner eyebrows.

6. Place critter in pot to form a foundation.

7. Place critter in sunlight and moisten when needed. The chia "hair" should sprout within a few days.

■ Bean Sprouts
[Makes about a cup of bean sprouts—enough for additions to several salads]

NOTE: Purchase beans from health food stores. Do not use packaged seeds for planting because the seeds may have been treated with chemicals. This process takes only a few days, and children can watch the beans germinate. Some kids like to eat the sprouts as a snack.

Materials

1/4 cup mung beans
very clean, clear plastic jar
very clean piece of cheesecloth,
     bigger than jar opening

rubber band
water

Process

1. Wash beans and remove any debris.

2. Pour beans into plastic jar and cover with plenty of water.

3. Place cheesecloth over jar opening and secure with rubber band.

4. Let seeds soak for up to 12 hours. Seeds will expand.

5. Drain water and place jar almost upside down so that extra moisture can escape. Keep away from sunlight.

6. Rinse and drain at least twice a day.

7. Watch seeds sprout. Decide the length of bean sprout for your taste. Rinse a last time and munch away!

■ Plants without Soil
[Makes 1 experiment]

NOTE: Hydroponics is the term for growing plants without soil. Children could research if fruits and vegetables could be grown without soil.

Materials

1 sponge

scissors

water

old aluminum tray or other device
    to catch extra water

mustard seeds

plastic wrap

spray bottle

Procedure

1. Cut sponge into an interesting shape.
2. Soak sponge in the water and then wring it out. The sponge should be moist but not dripping.
3. Place sponge on the aluminum tray.
4. Sprinkle mustard seeds on top of sponge.
5. Place tray in a sunny spot.
6. At night cover the tray with plastic wrap.
7. Fill spray bottle with water and mist sponge daily.
8. Within 2 weeks sprouts should appear.

■ Root Systems with and without Soil
[Makes 1 set]

NOTE: Children could see if the seeds without soil sprout at the same time as the seeds with soil.

Materials

2 large, clear plastic cups   water
sponge cut to closely fit the bottom   seeds
    of 1 plastic cup   potting soil

Procedure

1. Moisten sponge and put it on the bottom of one of the plastic cups.
2. Carefully arrange some seeds between the side of the plastic cup and the sponge.
3. Pour some of the potting soil into the other plastic cup.
4. Plant some seeds in the soil.
5. Moisten the soil.
6. Keep both the sponge and the potting soil damp.
7. Place both plastic cups in a sunny spot. Water as necessary.
8. Record the sprouting roots and leaves.

■ Miniature Terrariums
[Makes 1 plant]

NOTE: Children could make multiple bags. Some bags could be left open, and others could be closed. Is the rate of germination the same?

Materials

resealable plastic bag   about 10 bean seeds
1/2 cup soil   water

Procedure

1. Pour soil into the bag.
2. Place about 10 bean seeds into the soil.
3. Add a small amount of water and close the bag.
4. Place in a sunny area. The plastic bag should retain the water. Within about a week roots should appear.

12
Dried Flowers, Potpourri, and Pomanders

Dried Flowers

Flowers can be dried in many ways. Whatever the method, water must be removed. Dried flowers cannot be in sunlight, because sunlight will fade the flowers' colors. Dried flowers will usually last a year.

■ Dried Flowers Process 1 (Air-Drying Flowers, Leaves, and Herbs)

NOTE: Begin the drying process in early fall. Use some of the dried flowers and leaves to make holiday potpourri.

Materials

flowers, leaves, or herbs   rubber bands

Procedure

1. Pick flowers, leaves, and herbs at their prime.
2. Tie small bunches of the stems together with rubber bands.
3. Hang bunches petals down in a warm, dark, dry location for several weeks.

■ Dried Flowers Process 2 (Drying Flowers with Water)

NOTE: Thick-stemmed flowers such as zinnias dry well with this method.

Materials

flowers   water
vases

Procedure

1. Pour 1 inch of water into each vase. Stand flowers in vases.
2. Allow flowers to dry. The process may take as long as 2 weeks.

■ Dried Flowers Process 3 (Drying Herbs and Small Flowers in the Microwave Oven)

NOTE: Just about any herb or citrus rind can be dried via the microwave oven. However, very delicate flowers such as crocus will not dry well this way.

Materials

herbs or small but sturdy flowers
   such as daisies
paper towels

microwave oven
airtight container

Procedure

1. Place flowers or herbs on paper towel and place them in microwave oven.
2. Microwave at high setting for about 2 minutes. Stop the process and rearrange them on paper towel every 30 seconds.
3. Check to see if plants are dry. If not, microwave in periods of 30 seconds until dry.
4. Let cool and store in an airtight container.

■ Dried Flowers Process 4 (Glycerin)
[Makes 3 cups, enough to dry a large bunch of flowers]

NOTE: The glycerin is absorbed by the flowers and replaces the water. Glycerin can be purchased at a pharmacy.

Materials

2 cups hot water
1 cup glycerin
1 flower vase

fresh flowers
knife

Procedure

1. Combine glycerin and hot water in vase. Allow mixture to cool.
2. Split stems of the flowers with knife and place them in vase.
3. The flowers should be preserved in about 3 weeks.

■ Dried Flowers Process 5 (Borax–Cornmeal Mixture)
[Makes 3 cups drying mixture]

NOTE: This mixture can be used over and over. Borax can be bought in the detergents section of the grocery store.

Materials

1 cup borax	spoon
2 cups cornmeal	shoebox with lid
mixing bowl	fresh flowers

Procedure

1. Combine borax and cornmeal in bowl.
2. Spread about a half inch of mixture in bottom of shoebox.
3. Gently place flowers on top of mixture.
4. Cover flowers completely with more of mixture.
5. Place lid on box and store at room temperature for about a month.
6. Indicate the date that you put the flowers in and the probable date they will be ready.

■ Dried Flowers Process 6 (Borax)
[Makes 3 cups drying mixture]

NOTE: Borax can be bought in the detergents section of the grocery store.

Materials

3 cups borax	coffee can with lid
plastic bag	fresh flowers
plastic bag tie	soft paint brush

Procedure

1. Place plastic bag in coffee can.
2. Add enough borax so that about one inch of the material covers bottom of bag.
3. Put one flower in "face down." Cover with more borax.
4. Add more flowers and borax until bag is full.
5. Squeeze out air in bag and tie it shut.
6. Cover coffee can and put it somewhere in room temperature for about a month.
7. Make sure you write down the date you put the flowers in and the probable date they will be dry.
8. After a month remove flowers. Use paintbrush to gently brush away borax.

■ Dried Flowers Process 7 (Sand)
[Makes 3 quarts drying material]

NOTE: The sand can be used over and over again. Use this method if you cannot wait for weeks to go by.

Materials

1 old baking pan, 9 inches by
    12 inches by 2 inches
3 quarts dry sand
fresh flowers

oven
newspapers
soft paintbrush

Procedure

1. Cover bottom of baking pan with about one inch of sand.
2. Place the flowers "face up" in the sand.
3. Sift about an inch and a half sand over the flowers. Gently poke the sand around all the petals.
4. Bake at 200°F for about 2 hours.
5. Test to see if they are dry by gently pulling up one flower. Carefully bend a petal of the flower. If the petal breaks, the moisture is gone. The flowers are dry, and they should be removed from the oven. If the flower is still moist, bake 15 more minutes and then test again.
6. Remove flowers from sand and place on newspapers. Let them cool for about 2 hours.
7. Carefully brush off sand with soft paintbrush.

■ Dried Flowers Process 8 (Kitty Litter)
[Makes 1 quart drying material]

Note: Flowers can be dried in around a day with this method. The kitty litter can be used over and over.

Materials

1 quart kitty litter
microwave-safe container with lid

microwave oven
fresh flowers

Procedure

1. Place about an inch or so of kitty litter in bottom of microwave-safe container.
2. Place a few flowers on top of kitty litter. Surround flowers with more kitty litter.
3. Leave lid off and place the container in the microwave. Microwave from 1 to 3 minutes, depending on the type of flower. The thicker the flower, the longer the time should be.
4. Remove container from the microwave and loosely add lid.
5. Leave container alone for at least 18 hours so that flowers and kitty litter will slowly cool down and finish drying.
6. Remove dried flowers from container and shake off any extra kitty litter.

■ Pressing Flowers

NOTE: More delicate flowers, such as rose petals, press better than flowers with more bulbous bases, such as zinnias.

Materials

flowers
blotting paper (available at art supply stores or craft stores)

thin cardboard
heavy weight, such as old books

Procedure

1. Arrange a few flowers between sheets of blotting paper.
2. Cover with thin sheet of cardboard.
3. Place a heavy weight, such as a couple of old books, on top of the cardboard.
4. Leave in a warm, dry location for several weeks. Remove flowers when dry.

■ Making a Wreath with Dried Flowers

NOTE: Children can make small wreaths for holiday gifts. Having several adults present will make this activity easier and safer.

Materials

dried grapevines
large bucket of water
dried flowers
hot glue gun and glue

accents from nature, such as small pinecones or cinnamon sticks
wide strips of fabric or lengths of wide ribbon

Procedure

1. Soak grapevines in water until pliable.
2. Weave vines together to form wreath.
3. Tuck in many dried flowers.
4. Hot glue on extra touches, such as small pinecones or cinnamon sticks.
5. Tie fabric or ribbon into a large bow. Attach bow to wreath with more ribbon or with hot glue gun.

Potpourri

Children can be quite creative when they make potpourri. Basically, potpourri is any mix of dried flowers, dried leaves, dried fruit, spices, bark, and herbs. Children should choose dried flowers based on appearance, fragrance, and cost. Rose petals are nice but expensive. Petunia blossoms are cheaper but not so fragrant. Therefore, children might mix petunias and rose petals. Potpourri supplies can be purchased in craft stores.

A fixing agent is added so that the fragrances and color will last longer. Fixing agents include orrisroot, benzoin gum, balsam Tolu, and balsam Peru. Fixatives should not be consumed. Monitor their use by children. The potpourri can be made without fixatives, but the fragrances will not last long.

■ Strawberry Potpourri
[Makes 3 cups]

NOTE: Children can dry their own strawberries. The red and white colors make this potpourri a charming Valentine's Day project.

Materials

4 ounces dried, chopped strawberries
2 ounces dried strawberry leaves
8 ounces dried petunia blossoms, red and white
1/2 teaspoon cinnamon

2 tablespoons salt
1 ounce powdered benzoin gum (fixing agent)
plastic bag
decorative container

Procedure

1. Mix ingredients together in plastic bag.

2. Pour into decorative container.

■ Citrus Potpourri
[Makes 3 1/2 cups]

NOTE: The orange, yellow, and green colors brighten the potpourri.

Materials

3 ounces dried orange blossoms
2 ounces shredded, dried tangerine peel
1 ounce shredded, dried lemon peel
1 ounce shredded, dried orange peel
1 ounce shredded, dried lime peel

4 ounces dried bee balm blossoms
1 ounce powdered orrisroot (fixing agent)
plastic bag
decorative container

Procedure

1. Mix ingredients together in plastic bag.

2. Pour into decorative container.

■ Peppermint Potpourri
[Makes 2 1/2 cups]

NOTE: This potpourri has a "woodsy" appearance from the brown and gold chrysanthemum petals. Children could collect all the ingredients in the fall.

Materials

1 ounce dried peppermint leaves
1 ounce dried sweet woodruff
4 ounces dried chrysanthemum petals
1 ounce powdered balsam Tolu (fixing agent)

plastic bag
decorative container

Procedure

1. Mix ingredients together in plastic bag.
2. Pour into decorative container.

■ Spice Potpourri
[Makes 1 1/2 cups]

NOTE: Spice potpourri's ingredients can all be found at the grocery store. The aroma is heavenly!

Materials

1 teaspoon whole cloves
1 teaspoon whole allspice
1 teaspoon star anise
1 cinnamon stick
1 cup rock salt

1/2 vanilla bean, snapped
1 teaspoon vanilla extract
1/2 teaspoon almond extract
plastic bag
decorative container

Procedure

1. Mix ingredients together in plastic bag.
2. Pour into decorative container.

■ Lavender–Rose Potpourri
[Makes 6 cups]

NOTE: The various colors make this potpourri very attractive. Children could collect all the petals in the summer.

Materials

1 quart rose petals
2 cups lavender flowers
1 cup rosemary leaves
2 teaspoons whole cloves
6 tonka beans (fragrant seeds from *Dipteryxodorata*, a tree that grows in Central America; can be purchased through the Internet)

1/4 cup powdered orrisroot (fixing agent)
plastic bag
decorative container

Procedure

1. Mix ingredients together in plastic bag.

2. Pour into decorative container.

■ Old-Fashioned Pomanders

NOTE: Colonial children often made pomanders. Because citrus fruits were rare in colonial times, a pomander was a special gift. The orris root acts as a fixing agent. It can be left out of the process, but the fruit might mold sooner. Also, making and preserving a pomander takes about 2 weeks, so think ahead. Finally, the fruit will shrink as it dries.

Materials

thimble
oranges, lemons, or limes
For each piece of fruit:
100 whole cloves (approximately)
1 push pin
2 tablespoons ground cinnamon
1 teaspoon ground orrisroot (fixing
   agent)

2 plastic bags
17-inch square of netting
   (available at fabric stores)
1 yard ribbon (available
   at fabric stores)

Procedure

1. Poke a hole into citrus fruit with push pin. Push a whole clove into the hole in the fruit. Use the thimble to protect your fingers.

2. Repeat this process until the fruit is entirely covered with cloves.

3. Pour cinnamon into a plastic bag and orrisroot into other bag.

4. Place fruit in cinnamon bag and shake. Then move fruit to orrisroot bag and shake.

5. Place piece of fruit into square of netting.

6. Tie netting together with ribbon.

7. Hang for 2 weeks to dry.

13
Crystals

Crystals are easy to produce, and children like to see the changes. However, crystal solutions should not be disturbed (stirred, shaken, moved) after they are made. The two main ingredients for crystal growing are the chemical (solute) and water (solvent). A saturated solution is made by dissolving as much of the chemical, or crystal material, as possible in boiling water. When the solution cools, it becomes supersaturated. Crystals are by-products of this supersaturation.

Many crystals contain substances that should not be eaten. Also, many crystals are slow to grow; days can pass before changes can be observed.

■ Salt Crystals
[Makes 2 cups]

NOTE: Salt crystals are the easiest and cheapest crystals to grow. Crystals start to appear within 24 hours.

Chemical name: sodium chloride
Chemical formula: NaCl

Materials

1 cup salt
1 1/2 cups water
stove or heating element
pot and hot pads
mixing spoon

heat-resistant jar (e.g., canning jar)
string
paper clip, nail, or other small weight
pencil or stick longer than the diameter of the jar

Procedure

1. Heat water to boiling.
2. Gradually add some salt and stir. Keep solution boiling.
3. Add more salt and stir. Repeat until salt will no longer dissolve.
4. Remove pot from stove. Using hot pads, pour solution carefully into jar.
5. Cut a piece of string longer than the height of the jar. Tie one end to pencil or stick. Tie other end to paper clip or small weight.
6. Place pencil across top of jar so that string and weight dangle into salt solution.
7. Put jar in a place where it will not be disturbed. Soon crystals will grow on string.

■ Sugar Crystals
[Makes 2 cups]

NOTE: Sugar crystals are not always a sure thing. Sometimes a syrupy mess is the final product. Some people eat sugar crystals. However, because the solution has been exposed to the air and thus to germs and other contaminants, eating the crystals is not advised.

Chemical name: sucrose

Chemical formula: $C_{12}H_{22}O_{11}$

Materials

3 cups sugar
1 cup water
food coloring (optional)
stove or heating element
pot and hot pads
mixing spoon

heat-resistant jar (e.g., canning jar)
string
paper clip, nail, or other small weight
pencil or stick longer than the diameter of the jar
blanket

Procedure

1. Heat water to boiling.
2. Gradually add some sugar and stir. Keep solution boiling.
3. Add more sugar and stir. Repeat until sugar will no longer dissolve.
4. Remove pot from stove. Using hot pads, pour solution carefully into jar.
5. Cut a piece of string longer than the height of the jar. Tie one end to pencil or stick. Tie other end to paper clip or small weight.
6. Place pencil across top of jar so that string and weight dangle into sugar solution.
7. Put jar in a place where it will not be disturbed.
8. Add a bit of food coloring if desired.
9. Slow down cooling process as much as possible by wrapping jar in the blanket. Remove the blanket after a day. Soon crystals will grow on string.

■ Epsom Salt Crystals
[Makes 2 cups]

NOTE: Epsom salt crystals are needle-like and can grow quite large. They are also easy to grow.

Chemical name: magnesium sulfate
Chemical formula: $MgSO_4$

Materials
1 cup Epsom salt

1 1/2 cups water

stove or heating element

pot and hot pads

mixing spoon

heat-resistant jar (e.g., canning jar)

string

paper clip, nail, or other small weight

pencil or stick longer than the diameter of the jar

Procedure
1. Heat water to boiling.
2. Gradually add some Epsom salt and stir. Keep solution boiling.
3. Add more Epsom salt and stir. Repeat until Epsom salt will no longer dissolve.
4. Remove pot from stove. Using hot pads, pour solution carefully into jar.
5. Cut a piece of string longer than the height of the jar. Tie one end to pencil or stick. Tie other end to paper clip or small weight.
6. Place pencil across top of jar so that string and weight dangle into Epsom salt solution. Soon crystals will start to grow on the string.

■ Epsom Salt Frost
[Makes 2 cups]

NOTE: The liquid dishwashing detergent binds the Epsom salt to the glass. It also allows easy cleaning.

Materials

1 cup Epsom salt
1 1/2 cups water
stove or heating element
3 tablespoons liquid dishwashing
    detergent

pot and hot pads
mixing spoon
paintbrush
window or other glass surface

Procedure

1. Heat water to boiling.

2. Gradually add some Epsom salt and stir. Keep solution boiling.

3. Add more Epsom salt and stir. Repeat until Epsom salt will no longer dissolve.

4. Remove pot from stove.

5. Add liquid dishwashing detergent.

6. Let mixture cool.

7. "Paint" solution on window or glass with paintbrush. When solution dries, needle-like fan patterns will appear.

8. To clean, use soap and water to remove the Epsom salt frost.

■ Baking Soda Crystals
[Makes 1 cup]

NOTE: Baking soda crystals emerge within hours. They appear to take on fractal patterns on planar surfaces. Children "ooh" and "ahh" over these.

Chemical name: sodium bicarbonate

Chemical formula: $NaHCO_3$

Materials

1/3 cup baking soda	heat-resistant jar (e.g., canning jar)
1 cup water	string
stove or heating element	paper clip, nail, or other small weight
pot and hot pads	pencil or stick longer than the
mixing spoon	diameter of the jar

Procedure

1. Heat water to boiling.

2. Gradually add some baking soda and stir. Keep solution boiling.

3. Add more baking soda and stir. Repeat until baking soda will no longer dissolve.

4. Remove pot from stove. Using hot pads, pour solution carefully into jar.

5. Cut a piece of string longer than the height of the jar. Tie one end to pencil or stick. Tie other end to paper clip or small weight.

6. Place pencil across top of jar so that string and weight dangle into baking soda solution. Soon crystals will start to grow on the string.

■ Washing Soda Crystals

[Makes 1 1/2 cups]

NOTE: Washing soda can be found in the laundry products section of the grocery store.

Chemical name: sodium carbonate

Chemical formula: Na_2CO_3

Materials

2/3 cup washing soda
1 cup water
stove or heating element
pot and hot pads
mixing spoon

heat-resistant jar (e.g., canning jar)
string
paper clip, nail, or other small weight
pencil or stick longer than the diameter of the jar

Procedure

1. Heat water to boiling.
2. Gradually add some washing soda and stir. Keep solution boiling.
3. Add more washing soda and stir. Repeat until washing soda will no longer dissolve.
4. Remove pot from stove. Using hot pads, pour solution carefully into jar.
5. Cut a piece of string longer than the height of the jar. Tie one end to pencil or stick. Tie other end to paper clip or small weight.
6. Place pencil across top of jar so that string and weight dangle into washing soda solution. Soon crystals will start to grow on the string.

■ Borax Crystals
[Makes 1 1/2 cups]

NOTE: Borax crystals form against the sides of the jar as well as on the string. They are easy to make, and they start to appear within hours.

Chemical name: sodium tetraborate
Chemical formula: $Na_2B_4O_7$

Materials

2/3 cup borax*
1 cup water
stove or heating element
pot and hot pads
mixing spoon
heat-resistant jar
   (e.g., canning jar)

string
paper clip, nail, or other small weight
pencil or stick longer than the diameter of the jar

*Borax, found in the laundry products section of the grocery store, should not be eaten. Watch small children closely when they make this formula.

Procedure

1. Heat water to boiling.

2. Gradually add some borax and stir. Keep solution boiling.

3. Add more borax and stir. Repeat until borax will no longer dissolve.

4. Remove pot from stove. Using hot pads, pour solution carefully into jar.

5. Cut a piece of string longer than the height of the jar. Tie one end to pencil or stick. Tie other end to paper clip or small weight.

6. Place pencil across top of jar so that string and weight dangle into borax solution. Soon crystals will start to grow on the string.

■ Borax Stalagmites and Stalactites
[Makes 3 cups]

NOTE: These formations are not always a guarantee. Sometimes the string dries out, ending the process.

Materials

1 1/3 cups borax*
1 1/2 cups water
stove or heating element
pot and hot pads
mixing spoon
2 small, heat-resistant jars
(e.g., canning jars) of
the same size

several 12-inch pieces of
lightweight string
tray, big enough to hold both jars,
with 3 inches of space
between them

*Borax, found in the laundry products section of the grocery store, should not be eaten. Watch small children closely when they make this formula.

Procedure

1. Heat water to boiling.
2. Gradually add some borax and stir. Keep solution boiling.
3. Add more borax and stir. Repeat until borax will no longer dissolve.
4. Remove pot from stove. Using hot pads, take pot off stove.
5. Place jars on tray 3 inches apart.
6. Divide solution between jars.
7. Soak strings for 2 minutes in one of the jars.
8. Place one end of each string in the solution of one jar.
9. Place other end of each string in the solution of the other jar. The strings are thus suspended between jars.
10. Solution will begin to flow along strings. Some of the solution will drip from strings and solidify. Stalagmites and stalactites will form. Sometimes the two features will join and form a pillar of borax crystal material.

■ Cream of Tartar Crystals
[Makes 1 1/4 cups]

NOTE: Cream of tartar is one of the ingredients in baking powder. It is also used to make egg whites stiff for meringues and angel food cakes. Cream of tartar can be found in the spice section of the grocery store.

Chemical name: potassium bitartrate
Chemical formula: $KHC_4H_4O_6$

Materials

2/3 cup cream of tartar
1 cup water
stove or heating element
pot and hot pads
mixing spoon

heat-resistant jar (e.g., canning jar)
string
paper clip, nail, or other small weight
pencil or stick longer than the diameter of the jar

Procedure

1. Heat water to boiling.

2. Gradually add some cream of tartar and stir. Keep solution boiling.

3. Add more cream of tartar and stir. Repeat until cream of tartar will no longer dissolve.

4. Remove pot from stove. Using hot pads, pour solution carefully into jar.

5. Cut a piece of string longer than the height of the jar. Tie one end to pencil or stick. Tie other end to paper clip or small weight.

6. Place pencil across top of jar so that string and weight dangle into cream of tartar solution. Soon crystals will start to grow on the string.

■ Alum Crystals
[Makes 1 cup]

NOTE: Alum can be purchased at grocery stores in the spice section or at herb specialty shops. Alum is used to pickle cucumbers and other vegetables.

Chemical name: aluminum potassium sulfate
Chemical formula: $AlK(SO_4)_2$

Materials

2 ounces alum
1 cup water
stove or heating element
pot and hot pads
mixing spoon
2 heat-resistant jars
   (e.g., canning jars)

string
paper clip, nail, or other small weight
pencil or stick longer than the diameter of the jar

Procedure

1. Heat water to boiling.

2. Gradually add some alum and stir. Keep solution boiling.

3. Add more alum and stir. Repeat until alum will no longer dissolve.

4. Remove pan from stove. Using hot pads, pour solution carefully into one of the jars.

5. Let solution sit for 1 day. Several crystals should appear in the solution.

6. Pour the solution left in the jar into the other jar.

7. Remove the best of the alum crystals from the first jar. This will become the "seed crystal."

8. Cut a piece of string longer than the height of the jar. Tie one end to pencil or stick. Tie other end to seed crystal.

9. Place pencil across top of jar so that string and seed crystal dangle into alum solution.

10. Put jar in a place where it will not be disturbed. Soon crystals will grow around the seed crystal.

■ Crystal Critters and Shapes
[Makes 1 batch]

NOTE: Children can really explore some creative options. Again they can combine science with art.

Materials

batch of any of the above
   crystal solutions
pipe cleaners

large jar
pencil or stick longer than the
   diameter of the jar

Procedure

1. Pour batch of crystal solution into large jar.
2. Instead of suspending a string in the jar, take a pipe cleaner and make it into a star or snowflake.
3. Suspend the pipe cleaner over the pencil. The pipe cleaner shape should be submerged in the solution.
4. Crystals will start to grow on the pipe cleaner.
5. Children can then explore other pipe cleaner shapes, possibly a simple animal or snowflakes.

■ Suspended Sugar Crystals
[Makes 1 experiment]

NOTE: These crystals develop within a closed system. Evaporation is not part of the process.

Materials

1 tablespoon (1 envelope) unflavored
   gelatin
1 cup water
2 1/2 cups sugar
glass jar with lid

pot and hot pads
stove or heating element
mixing spoon
towel

Procedure

1. Combine unflavored gelatin and water in pot. Heat until water boils and then turn off heat.
2. Slowly stir in sugar, 1/4 cup at a time, until no more sugar will dissolve.
3. Pour liquid into glass jar. Leave any undissolved sugar at the bottom of the pot.
4. Screw lid on jar.
5. Wrap in towel (so that the mixture will cool slowly) and place in a spot where it will not be disturbed for weeks.

■ Salol Crystals
[Makes small amount of crystal]

NOTE: Salol can be purchased from science supply houses. The size of the crystals depends on the rapidity of the cooling process. The longer the cooling, the larger the crystals will be.

Chemical name: phenyl salicylate
Chemical Formula: $C_{13}H_{10}O_3$

Materials
1/2 oz. salol (phenyl salicylate)	shallow pan of warm water
1 small glass container	small, shallow aluminum pan

Procedure
1. Place salol in glass jar. Warm salol by placing jar in the shallow pan of warm water. Make sure no water enters jar.
2. After salol melts, pour into the small, shallow aluminum pan.
3. Crystals should form as the salol cools.

■ Sodium Thiosulfate Crystals
[Makes 1 batch]

NOTE: Sodium thiosulfate is also known at Photographer's Hypo, used to develop photographs. This experiment may not always work; even a bit of dust can cause the crystals not to form.

Chemical name: Sodium thiosulfate
Chemical formula: $Na_2S_2O_3$

Materials
1 box (16 ounces) sodium thiosulfate
1 glass or enamel pan
stove or heating element
1/4 cup water
mixing spoon
1 heat-proof jar, pint sized

old towel
hot pads
piece of cardboard slightly
    bigger than the mouth
    of the jar

Procedure
1. Heat water in glass or enamel pan.
2. Keep out a bit of the sodium thiosulfate. Slowly add the rest of sodium thiosulfate and stir constantly.
3. When all sodium thiosulfate is dissolved, carefully pour solution into jar.
4. Wrap towel around jar to slow down the cooling process.
5. Cover mouth of jar with piece of cardboard. This keeps out the dust.
6. Let solution cool to room temperature. This may take several hours.
7. Drop one speck of the sodium thiosulfate that was not dissolved into the solution. The single speck will cause the supersaturated solution to crystallize in just a few minutes.

■ Disappearing Crystals
[Makes 1 experiment]

NOTE: Soil Moist® crystals are polymers, which are long chains of atoms formed by joining identical groups of atoms. This particular polymer soaks up water to expand 300–400 times its first size.

Materials

about 15 Soil Moist® crystals
(available at nurseries)
2 glass jars, 1 with lid

distilled water
aluminum pie pan
string

Procedure

1. Place about 15 Soil Moist crystals in jar and fill with distilled water to about a 1/2 inch from the top.

2. Screw on lid and wait a few hours. The crystals should have really expanded.

3. Place several crystals on the aluminum pie plate. Choose one and gently tie a string around it.

4. Fill other jar with distilled water and lower crystal into it. Wait a few hours again. The crystal should disappear until someone pulls on the string and removes the crystal.

■ Crystal Garden 1 (Salt and Vinegar)
[Makes 1 crystal garden]

NOTE: This crystal garden goes through stages. First, buds of crystals appear on the coal. Fairly soon the coal is covered with well-developed clusters of crystals. After a few days the dish bottom displays large crystals. After the solution has evaporated, the crystals turn white and powdery. The whole process takes a week.

Materials

several pieces of charcoal, coal, or sponge
glass pie pan
petroleum jelly
1 cup basic "salt crystals" solution (see p. 131)

1/4 cup white vinegar
food coloring
mixing bowl
mixing spoon

Procedure

1. Place pieces of charcoal, coal, or sponge in glass pie pan.
2. Coat edge of glass pie pan with petroleum jelly to keep crystals in dish.
3. Mix salt crystal solution and vinegar together in mixing bowl. Pour over charcoal.
4. Dot surface with food coloring.
5. Let stand undisturbed for several days. Crystals will form on surface of charcoal and on dish bottom.

■ Crystal Garden 2 (Liquid Laundry Bluing and Ammonia)
[Makes 1 crystal garden]

NOTE: Children have been making crystal gardens for many generations. Now one of the main ingredients, laundry bluing, is hard to find. Sometimes you can find it in a grocery store, or you can order it on the Internet. The recipe will work without the bluing, but the results will not be as dramatic. Pieces of sponge work just as well as charcoal and are much cleaner.

Materials

several walnut-size pieces
   of charcoal, coal, or sponge
1 quart water
petroleum jelly
glass pie pan
6 tablespoons noniodized salt

6 tablespoons water
6 tablespoons liquid laundry bluing*
2 tablespoons ammonia*
food coloring
mixing bowl
mixing spoon

*The use of laundry bluing and ammonia must be monitored closely. Neither liquid should be ingested.

Procedure

1. Soak pieces of charcoal, coal, or sponge in 1 quart water for 20 minutes.
2. Coat edge of glass pie pan with petroleum jelly to keep crystals in dish.
3. Place pieces of charcoal, coal, or sponge in glass pie pan.
4. Mix all the other ingredients except food coloring in mixing bowl. Make sure salt is dissolved.
5. Pour solution over pieces of charcoal, coal, or sponge.
6. Dot surface with food coloring.
7. Do not move glass pie plate. Delicate crystals should start to appear in 20 minutes.

■ Homemade Geodes
[Makes 12]

NOTE: Children enjoy making these shiny gems. If kept dry, the crystals can last quite a long time.

Materials

1 empty, clean egg carton
plastic wrap
scissors
clean, empty egg shells

supersaturated solution of
salt or Epsom salt (see
p. 131)

Procedure

1. Cut plastic wrap into 12 pieces and line each of the 12 depressions in the egg carton.
2. Place an egg shell in each depression.
3. Pour some supersaturated solution into each of the egg shells.
4. Let mixture sit for several days. The egg shells should start to look like geodes.

■ Bath Crystals
[Makes 1 batch]

NOTE: These would make nice Mother's Day presents. Washing soda (sodium carbonate) can be found in the laundry products section of the grocery store.

Materials

1 cup washing soda crystals
plastic storage bag
food coloring
fragrance (essential oils or perfume)

small amount of water
eyedropper
wax paper
glass or plastic jar with lid

Procedure

1. Pour washing soda crystals in plastic bag.
2. Add a few drops of water and two drops of food coloring to the washing soda crystals.
3. Close bag and shake so that the color becomes uniform through the crystals.
4. Open bag. Add fragrance, close bag, and shake again.
5. Pour crystals onto the wax paper and allow to dry for about 10 minutes.
6. Pour crystals into the jar and seal.
7. To use, place a small amount of crystals into bath water.

14

Non-Newtonian Fluids, Slimes, and Goos

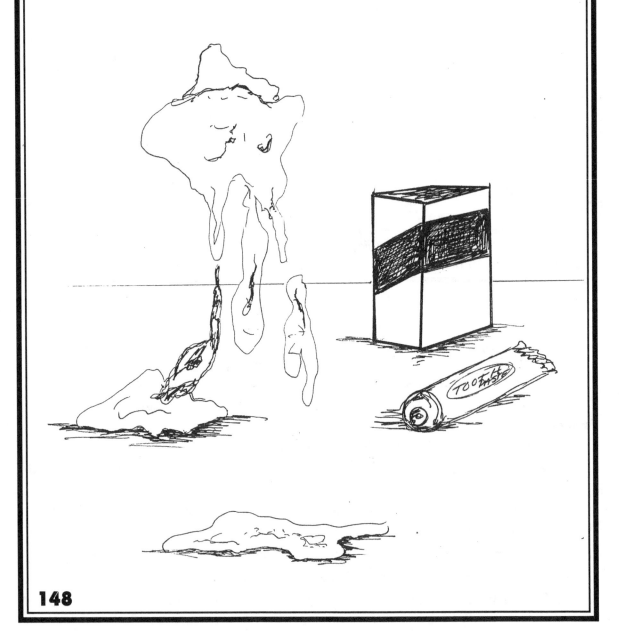

Non-Newtonian fluids fascinate children. These fluids have properties of both solids and liquids. They resemble fluids because they take the shape of their container. They resemble solids because they can maintain a definite shape. In the 1700s Isaac Newton developed a description of Newtonian fluids (fluids that "behave," meaning they take the shape of their container and they flow when the container is tilted) and Non-Newtonian fluids, substances that do not "behave." Ketchup is a Non-Newtonian fluid because it takes the shape of the bottle, but it also retains its own shape when it is removed from the bottle. Quicksand is also a Non-Newtonian fluid because it appears to be sand, but stress applied to it causes the water under the surface to quickly move the sand. Non-Newtonian fluids in this chapter can clog drains, so always dispose of the materials in a trashcan.

■ Non-Newtonian Fluid 1 (Cornstarch and Water)
[Makes 2 cups—enough for 4 children]

NOTE: Classic in its simplicity, this formula is a hit with children of all ages. In some ways this substance will act as a solid. At other times, it is a liquid. This will dry out if left exposed. It can be rejuvenated by adding a small amount of water. This is a fun activity after children read *Bartholomew and the Oobleck* by Dr. Seuss.

Materials

1 cup water
1 1/2 cups cornstarch

plastic, airtight container
spoon

Procedure

1. Pour water into container.

2. Slowly stir in cornstarch.

3. Children really like to mix this with their hands. Then they can begin stretching and squishing.

4. Store in airtight container.

■ Non-Newtonian Fluid 2 (White Glue and Liquid Starch)

[Makes 3/4 cup—enough for 1 child]

NOTE: It snaps! It rolls! It picks up print from newspaper! It is fun!

Materials

1/2 cup white glue   disposable container
1/4 cup liquid starch   wax paper
food coloring (optional)

Procedure

1. Mix white glue and liquid starch in disposable container.

2. Knead on wax paper until smooth.

3. If it is too sticky, add a bit more starch.

4. If it does not flow, add a bit more glue. Now children can have fun.

5. This non-Newtonian fluid does not last when stored. Use it the day you make it. Dispose in trash.

■ Non-Newtonian Fluid 3 (Borax and Polyvinyl Alcohol)

[Makes 2 cups—enough for 4 children]

NOTE: This product is similar to commercial "slimes."

Materials

1 cup borax*   plastic, airtight container
1 cup polyvinyl alcohol**   mixing spoon

*Borax, found in the laundry products section of the grocery store, should not be eaten. Watch small children closely when they make this formula.

**Polyvinyl alcohol, according to the manufacturer, is not toxic. However, children should not ingest it. Soap and water will remove polyvinyl alcohol from skin. It can be purchased from chemical supply houses.

Procedure

1. Combine borax and polyvinyl alcohol in disposable container. Stir for about 5 minutes. Children can try out the goo.

2. Store in airtight container.

■ Non-Newtonian Fluid 4 (Toothpaste, White Glue, and Cornstarch)
[Makes 1/2 cup—enough for 1 child]

NOTE: This concoction dries out after a while. However, it is fun to squish and squeeze.

Materials

2 teaspoons white toothpaste	small mixing bowl
4 teaspoons cornstarch	mixing spoon
2 teaspoons white glue	wax paper
1 teaspoon water	

Procedure

1. Combine toothpaste, cornstarch, and glue in bowl.
2. Add a bit of water and mix. Add a bit more water until mixture becomes a ball.
3. Place on wax paper, knead, roll, and enjoy.

■ Non-Newtonian Fluid 5 (White Glue and Borax)
[Makes 1 3/4 cups—enough for 3 children]

NOTE: The chemical reaction is quite thick and truly gooey. Children love it.

Materials

4 ounces white glue	1 teaspoon borax*
paper cup	disposable container (e.g., clean
1 1/2 cups water	butter tub)
plastic spoon	food coloring (optional)

*Borax, found in the laundry products section of the grocery store, should not be eaten. Watch small children closely when they make this formula.

Procedure

1. Pour glue into paper cup.
2. Add 1/2 cup water.
3. Mix borax and rest of water in disposable container.
4. Carefully pour glue mixture into borax mixture and mix. Add food coloring if desired.
5. Stir. Drain off extra liquid.
6. Knead until pliable. Children can squish and squeeze.
7. Store in a plastic bag. This dries out after a great deal of use.

■ Non-Newtonian Fluid 6 (Sculpting Material)
[Makes about 2 cups-enough for 2 children]

NOTE: Polystyrene beads, which can be bought at craft stores, are composed of mostly air. The beads provide a substrate for the goo. Children can use the goo to make sculptures. Projects can dry.

Materials

4 ounces white glue
paper cup
1 1/2 cups water
plastic spoon
1 teaspoon borax*

2 cups polystyrene beans
large disposable container
    (e.g., aluminum baking pan)
food coloring (optional)
wax paper

*Borax, found in the laundry products section of the grocery store, should not be eaten. Watch small children closely when they make this formula.

Procedure

1. Pour glue into paper cup.

2. Add 1/2 cup water.

3. Mix borax and rest of water in disposable container.

4. Carefully pour glue mixture into borax mixture and mix. Add food coloring if desired.

5. Slowly add polystyrene beads to mixture.

6. Remove some of mixture and place on wax paper. Mold into desired object. Let dry on wax paper.

7. Store any remaining mold mixture in refrigerator.

■ Gelatin Goo
[Makes 1 cup—enough for 2 children]

NOTE: Food coloring and/or flavorings could be added to this mixture.

Materials

1/2 cup boiling water
3 tablespoons (3 envelopes) unflavored gelatin
1/2 cup light corn syrup

heat-proof mixing bowl
mixing spoon
fork

Procedure

1. Pour boiling water into mixing bowl.
2. Add gelatin and let soften for several minutes.
3. Thoroughly combine gelatin and water.
4. Add light corn syrup and mix.
5. Use fork to pull out strands of goo.
6. This does not preserve well, so play with it the day it is made.
7. Do not pour this mixture down any drain.

■ Stretchers
[Makes 4 tablespoons—enough for 1 child]

NOTE: Children really enjoy playing with these.

Materials

2 tablespoons white glue
food coloring (optional)
1 tablespoon (1 envelope) unflavored
   gelatin
2 tablespoons boiling water

2 small mixing bowls
spoons
cookie cutter
wax paper

Procedure

1. Combine food coloring and white glue in one bowl.
2. Dissolve gelatin in boiling water in another bowl.
3. Combine the two mixtures and stir until batch thickens.
4. Place cookie cutter onto wax paper. Pour mixture into cookie cutter.
5. Allow to stand until mixture is firm.
6. Remove from cookie cutter and dry 1 hour on each side.
7. Children can now investigate their stretchers.
8. If allowed to dry, the stretchers become hard.

■ Fake Plastic
[Makes about 1 cup—enough for 2 children]

NOTE: This wonderful material is opaque.

Materials

1/4 cup water	kneading surface with extra flour
food coloring	2 small mixing bowls
1/4 cup white glue	mixing spoons
1/2 cup cornstarch	wax paper
1/2 cup flour	

Procedure

1. Combine liquids in 1 mixing bowl.

2. Combine flour and cornstarch in the other bowl.

3. Add flour/cornstarch mixture to liquid mixture and stir until dough becomes stiff.

4. Turn mixture out onto the kneading surface (with extra flour on it) and knead for several minutes.

5. Mold creations and place on wax paper.

■ Epsom Salt Goo
[Makes 1/4 cup—enough for 1 student]

NOTE: This goo is fun to make for a Halloween haunted house. Children are sometimes reluctant to touch the goo.

Materials

3 tablespoons white glue	disposable cup
1 1/2 teaspoons Epsom salt	disposable spoon
1 1/2 teaspoons water	paper towels

Procedure

1. Combine Epsom salt and water in disposable cup.

2. Add white glue.

3. Stir the mixture.

4. Pour it out on a paper towel to absorb the extra moisture.

5. Knead mixture until it becomes gooey.

■ Squeeze Goo
[Makes about 1 1/2 cups—enough for many projects]

NOTE: This goo makes nice accent touches to projects. It is fun to make and use.

Materials

1 cup flour	mixing bowl
1/4 cup salt	mixing spoon
1/4 cup sugar	squeeze bottle
3/4 cup water	wax paper
food coloring	

Procedure

1. Combine flour, salt, and sugar in mixing bowl.
2. Add water.
3. Add drops of food coloring until the desired shade is reached.
4. Pour mixture into the squeeze bottle.
5. Squeeze goo to form designs and patterns on wax paper. It will take a day or two to dry.

■ Gelatin Strings
[Makes about 20—enough for 10 children]

NOTE: The gelatin is a colloid, somewhat like a liquid and somewhat like a solid.

Materials

1 large package flavored gelatin	shallow pan
boiling water	plastic drinking straws
cold water	knife

Procedure

1. Prepare gelatin according to directions on package.
2. Pour mixture into shallow pan and place in refrigerator for about 30 minutes. The gelatin should be thick by then.
3. Sink straws into gelatin, and place pan back into the refrigerator. Leave it there for a day.
4. The next day use the knife to cut straws out of gelatin.
5. Pinch 1 end and begin to roll up the straw so that the gelatin strings pop out of the straws. Eat!

■ Guar Gum Slime
[Makes 8 batches of about 1/2 cup each]

NOTE: Guar gum can be bought at health food stores. It is a vegetable gum that is used to thicken foods and cosmetics.

Materials

2 teaspoons guar gum
1 tablespoon borax*
water
2 mixing bowls

2 mixing spoons
8 disposable cups
8 disposable spoons

*Borax, found in the laundry products section of the grocery store, should not be eaten. Watch small children closely when they make this formula.

Procedure

1. Combine borax and 1 cup water in one mixing bowl.
2. Pour 1 quart warm water into another mixing bowl. Add about 1 teaspoon guar gum and stir until dissolved.
3. Continue to add small amounts of guar gum to the warm water until all the guar gum is dissolved.
4. To make the slime, pour 1/2 cup of the guar gum solution into a disposable cup.
5. Add 1 teaspoon of the borax mixture, and stir together. Time to have fun!
6. The slime will last for a day or two. Then it reverts back to mostly water.

15
Bubble Solutions and Bubble Frames

Good bubble solutions must follow a number of rules. Most bubble solutions should be made days ahead, preferably even a week ahead, of use. Bubble solutions should be at room temperature. Use distilled water whenever possible. The size of the bubbles depends on such factors as humidity and air circulation. Also, the science seems to be that less is more where bubbles are concerned. In other words, the less soap used, the bigger the bubbles. If bubbles are used indoors, watch out for slippery floors by the end of the project.

■ Bubble Solution 1 (Glycerin)
[Makes 1 cup]

NOTE: Glycerin, available at most pharmacies, extends the life of the bubble.

Materials

2/3 cup water, preferably distilled
1/3 cup dishwashing liquid
1 teaspoon glycerin
green or yellow food coloring
   (optional)

mixing bowl
mixing spoon
airtight storage container

Procedure

1. Mix water, dishwashing liquid, and glycerin together. Add food coloring if desired.
2. Pour into airtight storage container. Allow it to age for a few days if possible.

■ Bubble Solution 2 (White Corn Syrup)
[Makes 1 1/4 cups]

NOTE: White corn syrup is cheaper and easier to obtain than glycerin. Like glycerin, corn syrup helps bubbles last longer.

Materials

1 cup water, preferably distilled
1/4 cup dishwashing liquid
1 tablespoon corn syrup

mixing bowl
mixing spoon
airtight storage container

Procedure

1. Mix water, dishwashing liquid, and corn syrup together.
2. Pour into airtight storage container. Allow it to age for a few days if possible.

■ Bubble Solution 3 (Gelatin)
[Makes 1 1/3 cups]

NOTE: Gelatin gives this solution a lumpy feeling. The bubbles are large and stable.

Materials

1 cup boiling water, preferably distilled

1 tablespoon (1 envelope) unflavored gelatin

1/3 cup dishwashing liquid

heat-proof mixing bowl

mixing spoon

airtight storage container

Procedure

1. In heat-proof bowl dissolve gelatin in boiling water.

2. Add dishwashing liquid.

3. Let cool and store in airtight storage container.

■ Bubble Solution 4 (Honey)
[Makes 1 1/3 cups]

NOTE: This solution needs hot water to properly dissolve honey. Honey is sometimes easier to obtain than either glycerin or white corn syrup.

Materials

1 cup hot water, preferably distilled

1/4 cup dishwashing liquid

3 tablespoons honey

heat-proof mixing bowl

mixing spoon

airtight storage container

Procedure

1. Mix hot water, dishwashing liquid, and honey together in heat-proof bowl.

2. Pour into airtight storage container. Allow it to age for a few days if possible.

■ Bubble Solution 5 (Shampoo)
[Makes 1 1/4 cups]

NOTE: Some shampoos work better than others. Built-in conditioners seem to help. Pert Plus® is a good choice. One advantage of this bubble solution is that it has a nice smell.

Materials
1/4 cup shampoo
1 cup water, preferably distilled
mixing bowl

mixing spoon
airtight storage container

Procedure
1. Mix shampoo and water together.
2. Pour into airtight storage container. Allow it to age for a few days if possible.

■ Bubble Solution 6 (Liquid Hand Soap)
[Makes 1 1/4 cups]

NOTE: This bubble solution is a better choice when you need a project right away. The bubbles seem to be tougher than those made with most other solutions. However, it will not keep for more than a day.

Materials
1/4 cup liquid hand soap (not
   antibacterial)
1 cup hot water, preferably distilled

heat-proof mixing bowl
mixing spoon

Procedure
1. Mix liquid hand soap and water together in heat-proof bowl.
2. Let cool and use immediately.

■ Bubble Solution 7 (Vegetable Oil)
[Makes 1 cup]

NOTE: The vegetable oil extends the life of the bubbles.

Materials

1/3 cup dishwashing liquid   mixing bowl
1 teaspoon vegetable oil   mixing spoon
2/3 cup water, preferably distilled   airtight storage container

Procedure

1. Combine dishwashing liquid, vegetable oil, and water in mixing bowl.

2. Pour into airtight storage container and allow it to age for at least 24 hours.

■ Bubble Solution 8 (Liquid Starch)
[Makes 1/2 cup]

NOTE: The mixture may at first be a bit thin, but it will thicken. It should be used as soon as possible.

Materials

1/2 cup liquid starch   mixing spoon
about 1 tablespoon dish soap   bubble wands
paper cups

Procedure

1. Pour liquid starch into a paper cup.

2. Add a generous dash of dish soap to the liquid starch and stir until combined.

3. Dip in the bubble wands and watch the results.

■ Liquid-Filled Bubbles
[Makes 1 batch]

NOTE: The liquid-filled bubbles actually have a layer of air surrounding them. This layer keeps the liquid inside from spreading into the surrounding liquid.

Materials

1/2 cup white corn syrup
water
1 teaspoon dishwashing liquid
4 drops food coloring

1/2 teaspoon salt
1 glass bowl
1 measuring cup
1 clean, empty squeeze bottle

Procedure

1. Pour white corn syrup into the bottom of the glass bowl. This layer will hopefully protect the liquid-filled bubbles as they fall.

2. Gently pour water into the bowl.

3. Add dishwashing liquid and gently swirl mixture to distribute.

4. Use measuring cup to scoop out some of the water–dishwashing liquid mixture.

5. Add food coloring to water in the measuring cup. The color will help you see the liquid-filled bubbles.

6. Add salt to the measuring cup to help the bubbles sink.

7. Pour salt-food coloring-water-dishwashing liquid into squeeze bottle and screw on the top.

8. Hold the opening of the squeeze bottle vertical to the bowl and gently squeeze.

9. Some of the liquid-filled bubbles will fall below the surface of the water in the bowl.

10. Children have fun watching the bubbles sink and then rise to the surface.

■ Tabletop Bubbles

NOTE: children's work areas have never been as clean as they will be after this experiment! Consider having children wear safety glasses, because bubbles can break close to their faces.

Materials

nonporous tabletops   wide straws
bubble solution   spoon

Procedure

1. Give each child a straw.
2. Pour about a tablespoon of bubble solution on table in front of each child.
3. Instruct child to put one end of straw into bubble solution and blow through the other end. Make sure bubble end of straw is really saturated with bubble solution.
4. With practice, children can make bubbles bigger than dinner plates. They can also make bubbles inside bubbles and bubble chains.
5. If a child is having difficulty getting a bubble started, swirl the solution around until a small bubble forms. Then the child can pierce the small bubble with the straw and blow slowly.

■ Bubble Frame 1 (Bell Wire)

NOTE: Plastic bubble wands found in commercial bubble solutions are just a beginning to this creative process. Bubble frames can be 3-dimensional or square-edged or both. No matter the shape of the frame, the bubbles will be spherical. Bell wire is best for these frames because it is easy to obtain and easy to cut. The plastic coating on bell wire protects children's fingers. This project is best done outside.

Materials

18 inches bell wire per child (available from hardware stores or science supply houses)

wire cutters
container of bubble solution

Procedure

1. Cut bell wire into 18-inch pieces, one for each child.

2. Have children make simple frames such as those found in commercial bubble bottles. Try out the frames.

3. Make frames of various shapes (square, triangular, 3-dimensional, etc.).

5. Place frames in container of bubble solution. Allow solution to really cover frames. Remove frames from solution and see what happens!

4. Record what works and what does not work.

5. Make a very large frame from wire. Have a contest to see who can make the biggest bubble.

■ Bubble Frame 2 (Straws and Strings)

NOTE: Science becomes fun with this rectangular but flexible bubble frame. Not only can children blow bubbles, but they can also manipulate the shape of the frame and thus the bubble film. Outdoors, children can make gigantic bubble frames by using dowels, long lengths of heavy string, and large plastic containers of bubble solution.

Materials

2 straws per child
1 40-inch length of string per child

shallow rectangular pans filled
with bubble solution

Procedure

1. Thread the string through one straw and then through the other.

2. Tie the ends of the string together so that a continuous loop of string has been formed.

3. Force the knot into one of the straws and pull the straws away from each other.

4. The resulting bubble frame is a rectangular shape with straws on the sides and strings on top and bottom.

5. Place bubble frames in rectangular pan of bubble solution. Allow solution to really soak into strings and straws.

6. Remove frames from solution and play. Twist, contort, and compress each frame. The bubble solution does amazing things.

7. If children get their hands very wet from the bubble solution, they can pass their hands through the bubble film on the frame.

16
Science Projects

Some people believe that science experiments require lots of test tubes and even more electronics. However, these science projects are inexpensive and use common ingredients. All of the following are all good fun.

■ Volcanic Action 1 (Strombolian Eruption)
[Makes 1]

NOTE: Stromboli is an island near Sicily. Its volcano, rising more than 3,000 feet above sea level, has been active for several millennia. It can erupt constantly for periods of up to several years. Stromboli an eruptions are seldom violent; lava and gases flow easily and are not pent up. In this experiment, the action is quick, producing an abundance of bubbles.

Materials

small, clean, empty plastic yogurt
    container
large pan
dirt, salt map mixture, or plaster
    of Paris

1 tablespoon baking soda
1 cup vinegar
red and green food coloring
    (optional)

Procedure

1. Place yogurt container in middle of pan. Make a "volcano shape" around the container with dirt, salt map mixture, or plaster of Paris. Make sure yogurt container forms the crater of the volcano. See p. 48 for specifics.

2. Pour baking soda into yogurt container.

3. Mix vinegar with food colorings (optional).

4. Pour vinegar into yogurt container. Very quickly the baking soda and vinegar will produce carbon dioxide, causing the volcano to erupt.

■ Volcanic Action 2 (Hawaiian Eruption)
[Makes 1]

NOTE: Hawaiian eruptions, named after the Hawaiian volcanoes, are fairly predictable. The lava often exits from several vents. Hawaiian eruptions are the least violent. The dishwashing liquid slows the eruption of this volcano, producing a cascading liquid that keeps its shape longer than the previous formula.

Materials

small, clean, empty plastic yogurt container	1 tablespoon baking soda
large pan	1 teaspoon dishwashing liquid
dirt, salt map mixture, or plaster of Paris	1 cup vinegar
	red and green food coloring (optional)

Procedure

1. Place yogurt container in middle of pan. Make a "volcano shape" around the container with dirt, salt map mixture, or plaster of Paris. Make sure yogurt container forms the crater of the volcano. See p. 48 for specifics.

2. Pour baking soda and dishwashing liquid into yogurt container.

3. Mix vinegar with food colorings (optional).

4. Pour vinegar into yogurt container. The dishwashing liquid delays contact between baking soda and vinegar. The overall reaction is longer and not so vigorously "volcanic" as the Strombolian Eruption.

■ Volcanic Action 3 (Vulcanian Eruption)
[Makes 1]

NOTE: Vulcanian eruptions get their name from Vulcano, an island near Italy. Inside this type of volcano, thick magma builds up inside a central vent. Ultimately gases increase pressure under the magma, and the magma explodes into dust and large pieces of debris. Effervescent antacid tablets contain, among other ingredients, sodium bicarbonate (baking soda) and citric acid. The addition of extra baking soda just keeps the process going longer. Although the volcanic reaction in this activity is not as fast as the Strombolian Eruption, the results are more explosive.

Materials

small, clean, empty plastic yogurt container
large pan
dirt, salt map mixture, or plaster of Paris

2 effervescent antacid tablets (e.g., Alka-Seltzer®)
1 teaspoon baking soda
red and green food coloring (optional)
1/2 cup water

Procedure

1. Place yogurt container in middle of pan. Make a "volcano shape" around the container with dirt, salt map mixture, or plaster of Paris. Make sure yogurt container forms the crater of the volcano. See p. 48 for specifics.

2. Place tablets, baking soda, and food coloring (optional) in yogurt container.

3. Add water. Watch the effects.

■ Volcanic Action 4 (Peléan Eruption)
[Makes 1]

NOTE: In 1902, Mount Pelée on Martinique erupted, killing more than 35,000 people. Peléan eruptions, named after Mount Pelée, are the most violent types of volcanic eruptions. The thick magma and gases clog a central vent in the volcano. Pressure builds until a terrifying explosion occurs—hot ash and dust cloud the atmosphere. Often, parts of the mountain portion of the volcano are destroyed. The steam from boiling water enhances the effects of the large, rolling bubbles. This is a favorite with children.

Materials

small, clean, empty plastic yogurt
   container
large pan
dirt, salt map mixture, or plaster
   of Paris

2 tablespoons baking soda
red and green food coloring
   (optional)
1/2 cup boiling water

Procedure

1. Place yogurt container in middle of pan. Make a "volcano shape" around the container with dirt, salt map mixture, or plaster of Paris. Make sure yogurt container forms the crater of the volcano. See p. 48 for specifics.

2. Pour baking soda into yogurt container.

3. Add boiling water. Stand back and watch!

■ Steam Eruption

[Makes 1]

NOTE: The interaction of the hydrogen peroxide and the yeast causes an exothermic reaction—heat is produced.

Materials

small, clean, empty plastic yogurt
    container
large pan
dirt, salt map mixture, or plaster
    of Paris

1 tablespoon yeast
1/2 cup hydrogen peroxide

Procedure

1. Place yogurt container in middle of pan. Make a "volcano shape" around the container with dirt, salt map mixture, or plaster of Paris. Make sure yogurt container forms the crater of the volcano. See p. 48 for specifics.
2. Pour hydrogen peroxide into yogurt container.
3. Quickly add yeast.
3. Steam and a sizzling noise should be produced.

■ Candy–Diet Soda Eruption

[Makes 1]

NOTE: This eruption must be done outside on a warm day! The eruption can be up to about 8 feet high. The person completing the last step must be a fast runner!

Materials

1 roll Mentos® candies
1 2-liter bottle diet soda at room
    temperature

1 sheet writing paper

Procedure

1. Roll the sheet of paper into a tube that is as big as the diameter of the candies.
2. Unwrap candies.
3. Find an open spot outside. Place bottle of diet soda on ground and remove cap.
4. Place 5 candies into tube. Place your hand over bottom of tube so that candies do not roll out.
5. Place tube in bottle opening. Let candies roll into bottle.
6. Run like crazy and watch the eruption!

■ Homemade Seismograph
[Makes 1]

NOTE: A seismograph measures the intensity of earthquakes. In this demonstration a child provides the earthquake.

Materials

1 round oatmeal box with lid	glue
string	about a cup of sand or other
roll of adding machine tape	filler
plastic, clear, disposable cup	scissors
pencil	

Procedure

1. Using scissors, cut a window on one side of oatmeal container from about 3 inches from the top all the way to the bottom. The width should be around 4 inches.

2. Cut the same type of window on opposite side of oatmeal box.

3. Thread adding machine tape through windows so that a portion of the tape rests on bottom of oatmeal container.

3. Poke a hole in top of oatmeal box.

4. Thread string through hole and tie a knot on top of box. String now hangs from inside of box.

5. Poke a hole in bottom of disposable cup. Put pencil through hole so that point of pencil is about an inch below bottom of cup. Glue pencil and cup together.

5. Tie hanging end of string to eraser of pencil so that string, cup, and pencil all hang from inside of oatmeal container. Pencil should just touch adding machine tape.

6. Fill cup with sand or filler to give weight to the cup/pencil portion.

7. The seismograph is complete. Now one child slowly pulls the adding machine tape through the seismograph. The pencil should leave a steady line on the tape.

8. If another child shakes the table while the other child pulls the tape, the pencil should leave all kinds of up and down lines!

■ Homemade Anemometer
[Makes 1]

NOTE: An anemometer measures wind speed. The faster the marked cup spins by, the stronger the wind.

Materials

1 wood block about 5 inches by 5
   inches by 2 inches
2 long nails
1 drinking straw
1 paper plate

3 small paper cups
3 brass paper fasteners
hammer
hole punch

Procedure

1. Drive 1 nail through the bottom of the wood block to form a base.
2. Place the end of the nail through the drinking straw and stand the base on a table.
3. Center the paper plate over the straw and poke the other nail through the plate and into the straw's opening. This should anchor the plate to the straw, but the plate should be able to spin.
4. Place the 3 paper cups on their sides on the plate so that they form an equilateral triangle. The lip of 1 cup should be close to the bottom of the cup ahead.
5. Attach the paper cups to the paper plate with brass fasteners. The cups should face the edge of the plate.
6. Make a mark on one of the cups.
7. Take the anemometer outside on a breezy day. Place the anemometer on a steady surface away from buildings. Allow the anemometer to spin, and count the number of times the marked cup rotates in 10 seconds.

■ Homemade Weathervane
[Makes 1]

NOTE: A weathervane measures wind direction.

Materials

1 wood block about 5 inches by 5 inches by 2 inches

2 long nails

1 drinking straw

1 piece of oaktag about 6 inches by 6 inches

glue

hammer

Procedure

1. Drive 1 nail through the bottom of the wood block to form a base.

2. Place the end of the nail through the drinking straw and stand the base on a table.

3. Cut two arrows from the oaktag. Each arrow should be about 6 inches by 2 inches.

4. Make a "sandwich" of the 2 arrows and place the other nail between them. Make sure the nail is both centered between them and perpendicular to them. Glue the arrows to the nail so that when the nail turns, the arrows move.

5. Place the nail into the straw opening. The arrows should be able to swing freely on the base.

6. Take the weather vane outside during a brisk wind. Does the arrow truly indicate the wind's direction?

■ Homemade Thermometer
[Makes 1]

NOTE: A homemade thermometer is not always successful, so be patient and flexible. Consider keeping a commercial thermometer nearby to see if the homemade version is at least rising or falling appropriately.

Materials

1 small, empty, clear plastic soda bottle
enough water to fill the bottle
3–4 drops food coloring

drinking straw
small wad of clay
bucket
hot water

Procedure

1. Fill plastic soda bottle with water.
2. Add food coloring so that the liquid can be made more visible.
3. Place some clay around the edge of the bottle opening.
4. Place some more clay around the middle of the straw.
5. Put the straw into the bottle and squeeze the clays together so that an airtight seal has been formed.
6. Place bottle in bucket and pour in some hot water. The water inside the bottle should expand, and some water should rise through the straw.

■ Homemade Hygrometer
[Makes 1]

NOTE: A hygrometer measures relative humidity. It compares the amount of moisture in the air to the amount of moisture air can hold at a certain temperature.

Materials

2 room thermometers that indicate
the same temperature
half-gallon paper milk carton (empty
and clean)
3 rubber bands

piece of cotton fabric about 5
inches square
scissors
water

Procedure

1. Cover bulb of one thermometer with the square cotton cloth. Some cloth should hang below thermometer. Attach thermometer and cloth with one rubber band.
2. Attach thermometers to one side of milk carton with remaining rubber bands.
3. With scissors, cut a hole in milk carton below thermometer that has cloth attached.
4. Push rest of cloth through hole so that cloth is inside milk carton.
5. Fill milk carton with water so that water level is above the cloth.
6. The thermometer with the cloth is called the wet bulb thermometer. The thermometer without the cloth is the dry thermometer.
7. Read and record the two temperatures. The drier the air, the more apart the readings should be. When precipitation falls, relative humidity is 100 percent.

■ Homemade Barometer (Water and Bottle)
[Makes 1]

NOTE: A barometer measures atmospheric pressure. Lower air pressure indicates stormy weather. Higher atmospheric pressure predicts good weather.

Materials

plastic cereal bowl   marker
empty and clean plastic soda bottle   water
masking tape

Procedure

1. Fill the plastic cereal bowl about 3/4 full with water.

2. Fill the plastic soda bottle about 3/4 full with water.

3. Put your thumb over the opening of the plastic soda bottle. Turn the bottle over and place it on in the cereal bowl. Remove your thumb and make sure soda bottle remains upright.

4. Place a piece of masking tape vertically along a side of the soda bottle. With the marker, mark the water level on the masking tape.

5. Watch the barometer over the next few hours or days. Mark when the water level in the soda bottle changes.

■ Homemade Barometer (Balloon and Jar)
[Makes 1]

NOTE: This barometer needs a wall.

Materials

wide-mouthed jar   scissors
balloon   a wall
2 rubber bands   1 sheet paper
straw   marker
masking tape

Procedure

1. Cut balloon in half. Stretch balloon half over mouth of jar.

2. Using rubber bands, attach balloon half over mouth of jar.

3. Tape straw to center of balloon/jar. Straw should hang over edge of jar.

4. Move barometer so that straw can touch the wall. Attach sheet of paper to wall so that straw touches sheet.

5 . With the marker, mark the straw level on the sheet of paper.

6. Watch the barometer over the next few hours or days. Mark when the straw level changes. What are the weather conditions when the straw level changes?

■ Homemade Rain Gauge
[Makes 1]

NOTE: Children could make several rain gauges and place them in various spots. Therefore, they could monitor rainfall and see if proximity to buildings or physical features changes the results. Children could also use math to chart results.

Materials

1 2-liter soda bottle with straight sides	water
	masking tape
scissors	thin marker
about a cup of sand or small rocks	ruler

Procedure

1. Remove cap from soda bottle. Cut top of soda bottle off where the slope ends. Retain the bottle top for step 5.
2. Fill the bottom of the bottle with sand or small rocks. The sand keeps the bottle from easily tipping over.
3. Pour water into the bottle so that water rises above the level of sand or small rocks. This point becomes "zero" on the gauge.
4. Place a piece of masking tape up the side of the bottle. The tape should start at the water level. Using the ruler, mark off every quarter inch.
5. Turn the bottle top (that was cut off) upside down and insert into the main portion of the bottle. The top acts as a funnel for the rain.
6. Place rain gauge outside and monitor results.

■ Ocean in a Bottle
[Makes 1]

NOTE: The "ocean" is caused by the fact that oil and water do not mix. Mineral oil can be purchased at a pharmacy.

Materials

1 empty, clear plastic 1-liter bottle
  with lid
1 1/2 cups mineral oil
small bowl

water
several drops blue food coloring
masking tape

Procedure

1. Pour mineral oil into bottle.
2. In bowl mix water and food coloring.
3. Pour enough blue water into bottle to fill it.
4. Screw on the lid and seal with tape.
5. Turn the bottle on its side and rock gently to watch the "ocean."

■ Miniature Tornado
[Makes 1]

NOTE: Children could find out how real tornadoes are generated. They could also find out what parts of the country get more tornadoes than other parts.

Materials

1 plastic soda bottle with a cap (16
  ounces is best)
enough water to fill the bottle

4 drops liquid dish detergent
2 teaspoons glitter or plastic
  confetti

Procedure

1. Fill bottle with cold water.
2. Add liquid dish detergent and glitter.
3. Screw on the cap.
4. Hold bottle upside down by the neck. Rotate the bottle quickly about 5 times in a clockwise motion. Stop and watch the miniature tornado in action inside the bottle.

■ Homemade Cloud 1 (Water and Match)
[Makes 1 cloud]

NOTE: Children could investigate how real clouds form.

Materials

empty, clear, plastic 2-liter soda bottle
   with cap
1 cup water
5 drops food coloring

match
flashlight
dark room

Procedure

1. Pour water and food coloring into soda bottle.
2. Light the match, blow it out, and immediately drop it into the soda bottle. Cap it.
3. Squeeze the bottle several times.
4. Take the bottle into the dark room and shine the flashlight through the sides of the bottle. You should see a cloud.

■ Homemade Cloud 2 (Water and Ice)
[Makes 1]

NOTE: Children can learn about the different kinds of clouds. They can observe the weather and make a daily chart of the clouds.

Materials

3 cups very hot water
clear glass measuring cup
old pair of panty hose

rubber band
10 ice cubes

Procedure

1. Pour hot water into the measuring cup.
2. Carefully touch the side of the measuring cup. When measuring cup becomes fairly hot, pour out all but about 1/2 cup of the water.
3. Stretch one section of the panty hose over the top of the measuring cup.
4. Fasten rubber band around the panty hose and the top of the measuring cup.
5. Place ice cubes on top of panty hose.
6. Water vapor will form small clouds.

■ Homemade Sunset
[Makes 1]

NOTE: The flashlight acts as the sun. The water is earth's atmosphere, and the milk is earth's air pollution. The light from the flashlight bounces around in the jar, and many of the colors of the spectrum stay within the jar. However, red light passes through the suspended droplets of milk, and so a red sunset is seen.

Materials

small glass jar with lid   flashlight
about 10 drops of whole milk   dark room
water

Procedure

1. Fill jar almost to the top with water.
2. Add milk.
3. Screw on lid and shake well.
4. Take jar and flashlight to a dark room.
5. Shine flashlight through one side of the jar. Look on the opposite side of the jar. A red sunset should appear.

■ Pennies of a Different Color

NOTE: In this activity, some of the copper from the pennies will appear on the nail. Some of the iron from the nail will deposit on the pennies.

Materials

1/4 cup vinegar   1 iron nail
3 teaspoons salt   small jar
5 pennies

Procedure

1. Combine vinegar and salt in jar.
2. Add pennies and nail.
3. Check in 2 hours. Some of the copper should be on the nail. Some of the iron should be on the pennies.

■ Return of the Pennies

NOTE: This activity works because ammonia is a base. It counteracts the effects of the vinegar, an acid. An interesting side effect is how the color of the ammonia changes.

Materials

Pennies from "Pennies of a Different Color" (see the preceding experiment)

1/2 cup ammonia*
small jar
paper towel

*The use of ammonia must be monitored carefully. Do not let children drink it, and watch that it does not splash on anyone.

Procedure

1. Place pennies from the previous experiment into the small jar.
2. Add enough ammonia to cover pennies.
3. Let sit several minutes.
4. Remove pennies and place on paper towel. The pennies will have returned to a copper color. The ammonia will be an interesting shade of blue.

■ Vanishing Color

NOTE: Bleach makes color disappear. Bleach contains sodium hypochlorite. When the oxygen in sodium hypochlorite mixes with food coloring, a new, colorless substance forms.

Materials

6 drops of any food coloring
1 cup water
small jar

eyedropper
small amount of bleach*

*Bleach should be handled very carefully. It can damage fabrics and harm skin.

Procedure

1. Mix food coloring and water in jar.
2. CAREFULLY add a drop of bleach.
3. Observe any color changes. Slowly add more drops of bleach until the color disappears.

■ Bouncing Egg
[Makes 1]

NOTE: Vinegar dissolves the calcium in the outer shell of a raw egg, leaving only the egg's inner membrane to keep the egg together. The egg can be gently bounced and squeezed. Keep plenty of paper towels on hand in case the egg breaks.

Materials
1 raw egg
white vinegar

container (with lid) big enough to hold egg

Procedure
1. Place egg in container.
2. Pour enough white vinegar into container to cover egg.
3. Place lid on container and let stand for a day.
4. After 24 hours, remove lid. Discard white vinegar.
5. Remove egg. The egg's hard outer shell has been dissolved by the vinegar. The egg will be squishy and rubbery.

■ Bending Bone
[Makes 1]

NOTE: As in the previous activity, the white vinegar dissolves the calcium in the bone. The bone becomes rubbery.

Materials
1 very clean chicken bone, preferably the thigh
white vinegar

container (with lid) big enough to hold bone

Procedure
1. Place bone in container.
2. Pour enough white vinegar into container to cover bone.
3. Place lid on container and let stand for a day.
4. After 24 hours, remove lid. Discard white vinegar.
5. Remove bone. Some of the bone's calcium has been dissolved by the vinegar. The bone will be bendable.

■ Testing for Starch

NOTE: This experiment is great as part of a unit on nutrition.

Materials

wax paper

different kinds of foods, including cut
potatoes, cut apples, orange segments, slices of bread, crackers,
and slices of cheese

tincture of iodine*

*Tincture of iodine is poisonous and flammable. Supervise its use carefully. It can
be purchased at pharmacies.

Procedure

1. Place foods on wax paper.
2. Place a drop of tincture of iodine onto each food.
3. The foods containing starch will turn purple where the tincture of iodine
   touches them.
4. Foods containing starch include bread, crackers, and potatoes. Children can
   record which foods contain starch.

■ Homemade Camera
[Makes 1 camera]

NOTE: The image should be upside down in this camera. Be patient; this project requires tweeking.

Materials

1 round, empty oatmeal box without lid

scissors

1 piece of aluminum foil about 3 inches by 3 inches

1 pin

1 sheet white tissue paper

masking tape

a small light source

dark room

Procedure

1. Use scissors to cut a hole about a 1/2 inch across in bottom of oatmeal box.
2. Cover the hole with the aluminum foil and secure with masking tape. The
   aluminum foil's shiny side should face the oatmeal box.
3. Make a small hole in the center of the aluminum foil with the pin.
4. Cover open end of box with a layer of tissue paper and secure with masking tape.
5. Turn on the small light source in the dark room. Point the pinhole toward the
   light. The small light source should be projected upside down on the tissue paper.

■ Easy Prism
[Makes 1]

NOTE: A prism breaks up visible light into its various parts.

Materials

jar
water

mirror small enough to fit into jar
white paper

Procedure

1. Fill jar with water.

2. Place mirror in jar at an angle.

3. Turn jar so that the mirror faces the sun.

4. Hold paper at an angle in front of the mirror. Move the paper until bands of color become distinct.

■ Amazing Water: Demonstration of Adhesion and Cohesion
[Makes 1 demonstration]

NOTE: This experiment works because of adhesion and cohesion. Water molecules like to stick to each other (cohesion). Water molecules also like to cling to other substances (adhesion). Children might want to practice this outside until they become good at it.

Materials

large measuring cup with pouring spout

another smaller glass container

water

several drops food coloring

piece of string about 36 inches long

Procedure

1. Fill measuring cup with water.

2. Add enough food coloring to make liquid very visible.

3. Soak string for about 30 seconds in water.

4. Tie 1 end of the string to the measuring cup handle and pull the string across the spout.

5. The other end of the string should be in the smaller glass container.

6. Hold the measuring cup in one hand and the string in the smaller glass container in the other hand.

7. Stretch the string as far as it will go and raise the measuring cup about 1 foot higher than the smaller glass container.

8. Slowly pour the liquid from the measuring cup.

9. Hopefully the liquid follows along the string and fills the smaller glass container. This takes some practice.

■ Layers of Liquid 1 (Honey and Vegetable Oil)
[Makes 1]

NOTE: Children could hypothesize about why this demonstration works. What would happen if they prepared another jar and put the honey in last?

Materials

1/2 cup honey or syrup
1/2 cup water
several drops red food coloring
1/2 cup vegetable oil

clear plastic pint jar
small bowl
spoon

Procedure

1. Pour honey into jar.

2. Combine food coloring and water in small bowl.

3. Pour food coloring/water mixture slowly on top of the honey. To slow down the process, place the spoon in the jar and trickle the water against the back of the spoon. The water should stay on top of the honey.

4. Slowly trickle the vegetable oil on top of the water. Children could record their observations.

■ Layers of Liquid 2 (Rubbing Alcohol)
[Makes 1]

NOTE: Children could experiment with the food coloring. Will the rubbing alcohol accept the color?

Materials

1/2 cup water
several drops red food coloring
1/2 cup rubbing alcohol*
clear plastic pint jar

small bowl
spoon
*Rubbing alcohol should not be
    consumed.

Procedure

1. Combine water and food coloring in small bowl. Pour mixture into jar.

2. Carefully pour rubbing alcohol on top of water. The two layers should not mix. Children could record their observations.

■ Temporary Magnet
[Makes 1]

NOTE: This project requires patience. The magnet will not be very strong, and it will not be permanent.

Materials
1 stainless steel needle   paperclips
1 strong bar magnet

Procedure
1. Stroke the stainless steel needle in one direction about 250 times with the bar magnet. The needle will be magnetized.
2. Try the new magnet out by placing it near a paperclip. What happens? Can the new magnet attract more than one paperclip?

■ Homemade Compass
[Makes 1]

NOTE: Orienteering is navigating between checkpoints on an outdoor course by using a compass. Children could make a compass good enough to use in orienteering and even create their own orienteering course.

Materials
3 needles   thread
bar magnet   pencil
index card   tape
scissors   jar with wide mouth

Procedure
1. Rub 2 needles about 250 times in the same direction from the eye to the point with the north pole of the magnet. Both needles are now magnets.
2. Cut the index card to a size 2 inches by 3 inches and fold lengthwise.
3. Unfold the index card. Tape a needle to each inside fold. Make sure the needles are parallel. Both points should face the same direction.
4. Refold the card.
5. Thread the third needle and poke through the fold from inside. Tie off the thread and leave at least 4 inches of thread free. Cut off the needle.
6. Tie the end of the thread around the middle of a pencil so that the card moves freely.
7. Place the card inside the jar and suspend the pencil across the top of the jar.
8. The card should rotate until it points north-south.
9. This magnet is temporary. Eventually the needles will lose their magnetism.

■ Periscope
[Makes 1]

NOTE: Submariners and spies use periscopes. Children could try to make telescoping periscopes, ones that change lengths.

Materials
2 clean, empty, half-gallon
    waxboard milk cartons
scissors

2 small mirrors
2 pieces of cardboard
tape

Procedure
1. Cut tops off of milk cartons.
2. Trim 1 piece of cardboard so that it will fit into 1 milk carton.
3. Tape a small mirror onto the cardboard.
4. Place the piece of cardboard with mirror facing out so that one cardboard edge rests on a carton bottom edge. The opposite cardboard edge rests at a 45 degree angle to the bottom of the milk carton. Tape the piece of cardboard in place.
5. Cut a small square away from the milk carton on the side opposite the one the cardboard rests on.
6. Repeat steps 2 through 5 with the other milk carton, cardboard, and mirror.
7. Join the 2 milk cartons together at their open ends. Make sure the mirrors are parallel to each other.
8. Tuck 1 carton slightly under the other so that they are fastened together. Tape the two milk cartons together.
9. Now face a door or other object. Look through one small opening of the periscope. Can you see the door or object? If so, the periscope is a success. If not, the mirrors may need to be adjusted.
10. See if the periscope can see around corners.

■ Screaming Cups
[Makes 1]

NOTE: The terrible sound is caused by the rough edges of the dental floss making contact with your fingers. This causes the string to vibrate, and the cup acts as a megaphone. Over time and use the dental floss loses its rough edges.

Materials

1 paper cup

1 toothpick

1 piece of ribbon waxed dental floss, 25 inches long

Procedure

1. Use toothpick to punch hole through bottom of cup.

2. Thread ribbon dental floss through hole.

3. Tie end of dental floss around the middle of the toothpick. The toothpick should be outside the cup's bottom.

4. Hold the cup in one hand. Pinch the dental floss with your other hand. Run your fingers along the dental floss. It should make a terrible sound.

5. Eventually the wax on the dental floss will become smooth, and there will be no terrible sound. Simply replace the dental floss to bring back the sound.

■ Simple Telephones
[Makes 1 set]

NOTE: Children could see if these phones work when the string goes around corners. Could they add more lines to the phones?

Materials

2 paper or plastic cups   scissors
10 feet of heavy duty string

Procedure

1. With scissors punch a small hole in bottom of each cup.
2. Push one end of string through hole from the bottom into the cup.
3. Tie a knot in the string so that the string cannot slip out of the bottom of the cup.
4. Do the same for the other cup.
5. Now the string connects both cups, the telephones.
6. Pull string tight.
7. One child should talk into the cup, and the other child should listen. Then they can change roles.

17
Geology Fun

Learning about rocks can be fun and interesting. Most of these activities let children make rocks and then eat them! Only the space rocks activity does not produce an edible product.

■ Conglomerate Cookies
[Makes 36 cookies]

NOTE: Make sure no one is allergic to nuts because this recipe uses peanut butter. Conglomerate rocks are composed of small pieces of mineral combined together through heat or by a cement-like substance. In this demonstration the sprinkles represent the smaller pieces, and the peanut butter is the cement.

Materials

1/4 cup shortening	lots of sprinkles
1/4 cup butter	nonstick cooking spray
1/2 cup peanut butter	mixing bowl
1/2 cup sugar	mixing spoon
1/2 cup brown sugar, packed	wax paper
1 large egg	cookie sheets
1 1/4 cups all-purpose flour	fork
3/4 teaspoon baking powder	refrigerator
3/4 teaspoon baking soda	oven
1/4 teaspoon salt	

Procedure

1. Combine shortening, butter, peanut butter, sugar, brown sugar, and egg in mixing bowl.
2. Add a portion of the flour and combine.
3. Add baking powder, baking soda, and salt. Again combine.
4. Add rest of flour and stir.
5. Cover bowl with wax paper and refrigerate for 2 hours.
6. Remove from refrigerator and roll pieces of dough the size of large marbles.
7. Mix in the sprinkles.
8. Spray cookie sheets with nonstick cooking spray.
9. Place balls of sprinkle-covered dough 3 inches apart on cookie sheets.
10. Flatten with fork tines.
11. Bake at 375°F for 10–12 minutes or until edges become a bit brown.

■ Volcanic Candy
[Makes about 24 pieces]

NOTE: Regular chocolate could be substituted for white chocolate. The red food coloring would then be unnecessary.

Materials

2 packages white chocolate (6 squares each)
red food coloring
microwave-safe bowl

microwave oven
mixing spoon
wax paper
cookie sheet

Procedure

1. Melt white chocolate in microwave-safe bowl (about 3 minutes).
2. Add food coloring to melted white chocolate.
3. Place wax paper on cookie sheet.
4. Drop mixture by spoonfuls onto wax paper.
5. Let harden—and eat!

■ Edible Conglomerate Rocks
[Makes about 24 rocks]

NOTE: Children are making crisped rice treats with a twist. The raisins, chocolate chips, and nuts represent rocks being forced together.

Materials

80 large marshmallows
7 tablespoons butter
12 cups crisped rice cereal
8 cups raisins, chocolate chips, nuts or any combination of these ingredients

large microwave-safe container
microwave oven
mixing spoon
nonstick cooking spray
small piece of wax paper for each student

Procedure

1. Combine marshmallows and butter in microwave-safe container. Microwave until marshmallows have melted.
2. Stir in crisped rice cereal.
3. Coat children's hands with nonstick cooking spray.
4. Give each child a small amount of the crisped rice cereal mixture.
5. Each child can then choose from the nuts, etc. to add to his/her mixture. These ingredients can be pressed into mixture to form the conglomerate rock.
6. Let "rocks" cool on wax paper before eating.

■ Sedimentary Layer Cake
[Makes about 10 slices of cake]

NOTE: Sedimentary rocks form as layers. Three-fourths of the earth's rock is sedimentary. This recipe allows children to see the layers. Children could add some sprinkles to the layers after the frosting has been applied.

Materials

1 cake mix and the required ingredients

2 containers of frosting

1/2 cup each of 4 toppings, such as toasted coconut, chocolate chips, sprinkles, or raisins

mixing bowl

mixing spoon

mixer

baking pans

length of dental floss

cake plate

oven

paper plates and plastic forks

Procedure

1. Prepare cake according to the box's instructions. Bake in two pans. Allow the layers to cool.

2. Remove the layers from pans and place on a work surface.

3. Cut each layer in half horizontally by using dental floss as a type of saw. Now four layers of cake are on work surface.

4. Place first layer on the plate and spread on some frosting. Add a topping.

5. Use the frosting to "glue" all the layers together, adding a topping between each of the other layers and on top of the last layer.

6. After the first piece is cut and removed, children can see the 4 layers of cake, the 4 layers of frosting, and the 4 layers of toppings.

■ Space Rocks
[Makes 10]

NOTE: Children could make these rocks and then create dioramas of the moon. They could color the liquid mixture red and make Mars rocks. Children could find out why they cannot make Jupiter or Saturn rocks. Children could also note the heat generated by the plaster of Paris and water.

Materials

3 cups plaster of Paris water
10 resealable plastic sandwich bags hammer

Procedure

1. Pour some plaster of Paris into each bag.

2. Add enough water to make a thick paste.

3. Seal the bags and squish the bags so that the water and plaster of Paris become thoroughly mixed. Let plaster of Paris harden for several hours.

4. Once the plaster of Paris is hard, hit each bag with a hammer to break the solid into rocks.

18
Making Musical Instruments

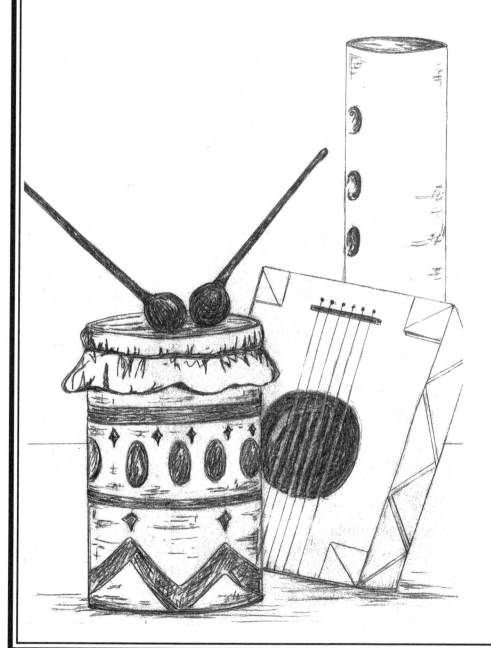

Most musical instruments create sound by blowing through a tube (e.g., a trumpet), vibrating strings (e.g., a violin) or hitting something (like a drum). Children learn a great deal about music when they make their own instruments, and they also learn about the science of sound.

■ Water Glass Musical Instruments
[Makes 1 set]

NOTE: Children could note that the more the water, the lower the sound.

Materials

8 glasses, each at least 6 inches tall   water
ruler   metal spoon

Procedure

1. Line up 8 glasses.
2. Fill first glass to the top with water to produce the low C note.
3. Use ruler to fill the next glass 8/9 full to produce the D note.
4. Use ruler to fill the next glass 4/5 full to produce the E note.
5. Fill next glass 3/4 full to produce the F note.
6. Fill next glass 2/3 full to produce the G note.
7. Use ruler to fill the next glass 3/5 full to produce the A note.
8. Fill next glass 8/15 full to produce the B note.
9. Fill last glass halfway to produce the high C note.
10. Gently tap each glass with metal spoon and begin to make music.

■ Panpipes
[Makes 1 set]

NOTE: Many cultures, from ancient Egypt to present-day Peru, use versions of the panpipes.

Materials

about 40 inches PVC pipe, 1/2-inch in diameter	hacksaw
2 craft sticks	sandpaper
duct tape	5 corks that will fit into the 1/2-inch PVC pipe

Procedure

1. Cut 5 lengths of pipe so that each pipe will play one note. The following indicates the lengths to be cut and the corresponding notes that can be played.
2. Cut length of pipe 9 1/2 inches to produce a G note.
3. Cut length of pipe 8 1/2 inches to produce an A note.
4. Cut length of pipe 7 1/2 inches to produce a B note.
5. Cut length of pipe 7 inches to produce a D note.
6. Cut length of pipe 6 inches to produce an E note.
7. Sandpaper pipe edges so that there are no sharp edges.
8. Line up the 5 pieces from biggest to smallest and so that the tops are all equal.
9. Sandwich the tops between 2 craft sticks and bind together tightly with duct tape.
10. Place corks in the other ends of panpipes.
11. Adjust the pitch of each pipe by slightly raising or lowering cork.
12. Play panpipes by blowing across the tops of the tubes.

■ Boom Pipes
[Makes 1 set]

NOTE: These pipes, providing a wonderful, resonant sound, can make school concerts special! They can be played inside or outside. PVC pipe usually comes in 10 foot lengths.

Materials

6 lengths of 4-inch PVC thin-wall
   drain pipe
14 end caps for the PVC pipe

14 carpet squares
hacksaw

Procedure

1. Cut 14 lengths of pipe so that each pipe will play 1 note. The following indicates the lengths to be cut and the corresponding notes that can be played.
2. Cut length of pipe 19 inches to produce an E note.
3. Cut length of pipe 22 inches to produce a D note.
4. Cut length of pipe 24 3/4 inches to produce a C note.
5. Cut length of pipe 26 1/4 inches to produce a B note.
6. Cut length of pipe 29 1/2 inches to produce an A note.
7. Cut length of pipe 33 1/2 inches to produce a G note.
8. Cut length of pipe 35 1/2 inches to produce an F# note.
9. Cut length of pipe 37 3/4 inches to produce an F note.
10. Cut length of pipe 40 1/2 inches to produce an E note.
11. Cut length of pipe 45 inches to produce a D note.
12. Cut length of pipe 50 1/2 inches to produce a C note.
13. Cut length of pipe 55 inches to produce a B note.
14. Cut length of pipe 61 inches to produce an A note.
15. Cut length of pipe 68 1/4 inches to produce a G note.
16. Cover 1 end of each pipe with an end cap.
17. Play the pipes by booming a pipe (end cap down) against a carpet square.
18. If a pipe is a bit flat, cut off 1/4 inch.
19. If a pipe is too sharp, add a bit of water.

■ Kazoo
[Makes 1]

NOTE: Music is created when the humming causes the wax paper to vibrate. Children could decorate the tubing with markers or paint.

Materials
1 cardboard tube from paper towels
1 sheet wax paper about 8 inches by
    8 inches

1 rubber band
1 sharpened pencil

Procedure
1. Place wax paper over one end of tube and fasten with the rubber band. Make sure wax paper is flat over the opening.
2. With pencil point, make a small puncture in the tube about 2 inches from the wax paper end.
3. To play, hum through the open end of the kazoo. See what happens when the hole is covered.

■ Straw Trombone
[Makes 1]

NOTE: Moving a real trombone's slide changes the pitch of the sound. In this trombone, moving the straw up or down in the bottle's water changes the sound.

Materials
1 empty soda bottle
1 straw (thicker is better)

enough water to fill soda bottle
    about 3/4

Procedure
1. Pour water into soda bottle.
2. Place straw in water.
3. Blow across the end of the straw that is not in the water. You can make a sound!
4. Move the straw deeper in the water to produce deeper sounds. Conversely, pull the straw up to produce higher notes.

■ Straw Oboe
[Makes 1]

NOTE: This oboe is a very quiet! You have to listen closely for the sound.

Materials

1 straw   scissors

Procedure

1. Flatten one end of straw, the last 3/4 of an inch.

2. Using scissors, trim corners off the same end you flattened.

3. Use scissors again to drill 3 small holes equally spaced along a side of the straw.

4. Blow through the straw to produce sound. Cover any combination of the 3 holes to produce different notes.

■ Tambourines
[Makes 2]

NOTE: Children could try different fillings inside the paper plates. Can different sounds be made? Children could also decorate the plates.

Materials

4 paper plates   stapler and staples
about 1/2 cup dried large beans,   masking tape
    beads, or uncooked pasta.

Procedure

1. Place 2 plates up on the table.

2. Pour about 1/4 cup beans, beads, or uncooked pasta into each plate.

3. Cover each plate with another plate so that the rims touch.

4. Staple the sets of plates together.

5. Completely seal the edges together with tape.

6. Shake or hit tambourines to make music.

■ Wood Stick Tambourine
[Makes 1 set]

NOTE: The number of bottle caps can be changed.

Materials

1 strip of wood about 2 inches by 2 inches by 9 inches
12 metal bottle caps

12 nails with wide heads about 3 inches long
hammer

Procedure

1. Make 4 stacks of bottle caps.

2. Place each stack on the strip of wood so that each stack is about 1/2 inch away from the next stack.

3. Hammer a nail through each stack so that the bottle caps are loosely fastened to the wood. The bottle caps should be able to move along the length of the nail.

4. Play the wood stick tambourine by shaking it or by gently banging it against another surface.

■ Rubber Band Guitar
[Makes 1]

NOTE: The frame for the rubber bands must be very sturdy. Children can investigate how different rubber bands can produce different pitches.

Materials

1 square metal tin about 6 inches by 6 inches (no lid is necessary)

about 6 big rubber bands of varying widths

Procedure

1. Place the rubber bands around the top and bottom of the metal tin. Make sure the rubber bands are parallel to each other.

2. Twang away!

■ Sand Blocks
[Makes 1 set]

NOTE: Different grades of sandpaper can make different sounds. This could lead to some fun experiments.

Materials

2 blocks of wood about 5 inches by 5
    inches by 2 inches
sandpaper
scissors

thumbtacks
2 empty spools
glue

Procedure

1. Cut a piece of sandpaper so that it covers the bottom and two sides of one of the blocks.
2. Thumbtack the sand paper to the block on two sides.
3. Repeat the procedure with the other block.
4. Glue a spool to the top (the large side not covered with sandpaper) of each block. The spools provide handles for the sand blocks.
5. Allow glue to dry.
6. Play the sand blocks by brushing them against one another.

■ Drums
[Makes 1]

NOTE: Just about every culture has created drums. Children could research particular types of drums.

Materials

empty oatmeal cartons or empty cof-
    fee cans with lids

2 pencils
masking tape

Procedure

1. Wrap enough masking tape around 1 end of each pencil so that small balls are formed. These become the heads of the drumsticks.
2. The oatmeal cartons or coffee cans become the drums.
3. Hit the oatmeal cartons or coffee cans with the drumsticks.
4. Can different pitches be created?

■ Steel Drums
[Makes 1 set]

NOTE: Real steel drums, popular in the Caribbean region, are made from cut oil drums that are pounded and tuned. Empty food tins can be substituted for the steel drums. Small tins make high notes, and big tins make low notes.

Materials

3 empty, clean food tins (cookie or popcorn tins work well)

2 unsharpened pencils
about 6 rubber bands

Procedure

1. The empty food tins become the drums. The tins could be painted or covered with pretty decorations.

2. Wrap the rubber bands around the unsharpened pencils to make the drumsticks.

3. Hit inside of tins with drumsticks. Have a good time!

19
Invisible Inks

Invisible inks have been around for hundreds of years. Most are chemical reactions. For example, iodine reacts to the presence of starch. Some invisible inks depend on a base reacting to an acid.

■ Invisible Ink 1 (Juice Invisible Ink)
[Makes 1/2 cup]

NOTE: Children enjoy experimenting with different solutions. The liquid materials are quite safe. Children should be supervised when they use the light bulb. A warm iron, monitored carefully, can also bring forth messages.

Materials

fine paintbrush

paper

1/2 cup of any of the following liquids: milk, lemon juice, grapefruit juice, orange juice, apple juice, sugar–water solution, clear soda such as ginger ale

lamp with bulb

Procedure

1. Dip brush into liquid.

2. Write a message on paper. The message should disappear as liquid dries.

3. To retrieve message, warm paper over light bulb.

■ Invisible Ink 2 (Vegetable Oil and Ammonia Invisible Ink)
[Makes 1 1/2 cups]

NOTE: This formula requires no heat. Many children can do the project at the same time. Just watch the ammonia.

Materials

1 teaspoon vegetable oil
5 tablespoons ammonia*
1 1/4 cups water
small container

mixing spoon
fine art brush
paper

*The use of ammonia must be monitored carefully. Do not let children drink it, and watch that it does not splash on anyone.

Procedure

1. Mix vegetable oil and ammonia in small container.
2. Add water carefully and stir.
3. Use brush to write message on paper. The message will disappear as paper dries.
4. To see the message, dip paper in water. The message can appear and disappear quite a few times.

■ Invisible Ink 3 (Citric Acid Invisible Ink)
[Makes 1 cup]

NOTE: With this ink, the message becomes visible when the paper background is stained blue by the tincture of iodine. The message itself does not change color because the vitamin C in citric acid combines with the iodine to form a colorless compound.

Materials

15 drops tincture of iodine*	2 small disposable bowls
3/4 cup water	paper
2 tablespoons citric acid (available at grocery stores in canning section)	cotton swabs
	rubber gloves

*Tincture of iodine is poisonous and flammable. Supervise its use carefully. It can be purchased at pharmacies.

Procedure

1. Combine tincture of iodine with 1/2 cup water in a bowl.
2. Combine citric acid and remaining 1/4 cup water in other bowl.
3. Dip cotton swab into citric acid mixture and write message on paper.
4. Allow message to dry.
5. Put on rubber gloves. Place message into bowl containing tincture of iodine and water. The message will reappear.

■ Invisible Ink 4 (Table Salt Invisible Ink)
[Makes 1/4 cup]

NOTE: This method is quite safe and requires no heat. However, it needs a day for the salt to sufficiently dry.

Materials

2 tablespoons salt	mixing spoon
1/4 cup water	fine paintbrush
small bowl	dark-colored construction paper

Procedure

1. Mix salt and water in small bowl.
2. Dip brush in salt water solution and write the first letter or two of message.
3. Dip brush again into solution and write another few letters.
4. Keep dipping and writing until message is complete.
5. Allow paper to dry overnight. Message should appear by the next morning.

■ Invisible Ink 5 (Wax Invisible Ink)
[Makes 1]

NOTE: Emergency candles can be purchased at hardware stores. Old candles with not much color could be substituted. This project is an example of a resist. The wax resists the water in the watercolors.

Materials
1 white emergency candle or white crayons
1 piece white construction paper

1 set watercolors and paintbrush
water

Procedure
1. Write a message on the construction paper with the candle or white crayons. The more wax used, the more successful the product will be.

2. To reveal the message, apply watercolor paints on top of paper. The wax should resist the watercolors, and the message will appear.

■ Invisible Ink 6 (Cornstarch Invisible Ink)
[Makes 1/4 cup]

NOTE: This process works because the tincture of iodine reacts to the starch in the cornstarch.

Materials
1 teaspoon cornstarch
1/2 cup water
microwave-safe bowl
microwave oven
mixing spoon

10 drops tincture of iodine*
disposable container
paper
2 swabs

*Tincture of iodine is poisonous and flammable. Supervise its use carefully. It can be purchased at pharmacies.

Procedure
1. Combine cornstarch and 1/4 cup water in microwave-safe bowl.

2. Microwave for about 30 seconds to thicken mixture.

3. Write message on paper with swab and let paper dry.

4. Combine iodine and 1/4 cup water in the disposable container.

5. With the second swab, paint the paper with the iodine mixture. The message will appear in dark blue or purple.

■ Invisible Ink 7 (Baking Soda Invisible Ink)
[Makes about 1/4 cup]

NOTE: The grape juice concentrate contains acids. These acids react to the base in the baking soda.

Materials

1/4 cup baking soda
1/4 cup water
small bowl
mixing spoon

1/4 cup grape juice concentrate
2 swabs
paper

Procedure

1. Combine baking soda and water in small bowl.

2. Use a swab to write a message with mixture on paper. Wait for the paper to dry.

3. Swab paper with grape juice concentrate. The message should appear.

■ Invisible Ink 8 (Vinegar–Red Cabbage Invisible Ink)
[Makes 1/4 cup vinegar solution and 1 cup indicator solution]

NOTE: The red cabbage water acts as an indicator and reacts to acidic vinegar.

Materials

1/4 cup white vinegar
1/2 head red cabbage
1 cup water
knife
pot

stove or heating element
mixing spoon
paper
2 swabs

Procedure

1. Use a swab dipped in white vinegar to write message on paper. Let paper dry.

2. Cut cabbage into small pieces and place in pot.

3. Add water and cook over low heat until water turns deep red or purple. Discard cabbage and keep the liquid.

4. Dip other swab in red cabbage liquid and paint paper. The message should stand out.

■ Invisible Ink 9 (Phenolphthalein Invisible Ink)

[Makes 1/4 cup phenolphthalein solution and 1/4 cup washing soda solution]

NOTE: The phenolphthalein acts as an indicator and reacts to the base in the washing soda. Washing soda (sodium carbonate) is found in the laundry products section of the grocery store.

Materials

3 laxative tablets containing phenolphthalein
1/2 cup water
3 tablespoons washing soda

2 small containers
mixing spoons
2 swabs
paper

Procedure

1. Place laxative tablets in one of the small containers. Crush tablets with back of spoon.
2. Add 1/4 water and mix.
3. Combine washing soda and 1/4 cup water in other small container.
4. Dip a swab in liquid from the phenolphthalein and write message on the paper. Let paper dry.
5. Dip the other swab in washing soda mixture and cover paper. The message should become light pink.

■ Invisible Ink 10 (Salt–Graphite Invisible Ink)

[Makes 1/4 cup solution]

NOTE: The salt and the graphite from the pencil do not get along.

Materials

3 tablespoons salt
1/4 cup water
small container
mixing spoon

1 swab
1 pencil
paper

Procedure

1. Combine salt and water in container. Stir until salt dissolves.
2. Dip swab in saltwater solution and write message on paper. Let paper dry.
3. Rub over the area with the graphite from the pencil. The message should appear.

20

Activities about Other Times and Other Cultures

Every culture has unique characteristics. However, many similarities exist between cultures. Most of the great ancient cultures flourished in warm regions with good farmland near rivers. Almost every culture developed advanced farming techniques and extensive food storage facilities. Almost every culture was concerned about the afterlife.

■ Mummifying Material
[Makes 2 1/2 cups material]

NOTE: The ancient Egyptians lived in a narrow band of land along the Nile River. They had to deal with sand, heat, and aridity. They built their homes from clay bricks, and they built their pyramids from stones transported hundreds of miles down the Nile. They devoted their lives in preparation for the afterlife. The ancient Egyptians were famous for their mummifying techniques. The mummification of a piece of fruit allows children to understand the mummification process. Washing soda (sodium carbonate) is located in the laundry products section of the grocery store.

Materials

1 cup baking soda
1 cup washing soda
1/2 cup iodized salt
1 apple or pear, cut in fourths

1 disposable container large enough
    to hold about 3 cups of material
mixing spoon

Procedure

1. Combine baking soda, washing soda, and salt in container.

2. Bury apple or pear pieces in mixture.

3. Let stand for 10 days and remove the dried apple or pear. Check to see how much it has desiccated.

■ Fake Papyrus
[Makes enough for 1 project]

NOTE: This project requires patience and luck. The Egyptians made papyrus writing material from papyrus reeds. In fact, etymologists trace the root of the word "paper" back to papyrus. The Egyptians did not weave the strips of papyrus, but they placed layers at 90-degree angles and pounded the fibers together.

Materials

3 stalks rhubarb	hammer
vegetable peeler	lots of newspaper
old cookie sheet with lip	heavy weights

Procedure

1. Place several layers of newspaper on old cookie sheet.
2. With the vegetable peeler, peel several length-wise strips of rhubarb stalks.
3. Lay about five strips of rhubarb stalk side by side on the newspaper.
4. Turn the cookie sheet 90 degrees. Place about five strips of rhubarb side by side on top of the first layer. The second layer should be perpendicular to the first layer.
5. Place two sheets of newspaper on top of the rhubarb.
6. Gently pound the rhubarb–newspaper layers with the hammer.
7. Remove the top layer of newspapers and add a new layer of papers.
8. Place several heavy weights on top of the layers.
9. Replace newspaper when moisture seeps through.
10. After two days remove the newspaper. The papyrus is one sheet, and it should resemble paper.

■ Sugar Cube Pyramid
[Makes 1 pyramid]

NOTE: The ancient Egyptians built huge pyramids as tombs for their pharaohs. This sugar cube pyramid is an easy way to show how pyramids were made. Children should not eat the icing because it contains raw egg white.

Materials

1 egg white
1/8 teaspoon cream of tartar
1 1/2 cups confectioners' sugar
at least 385 sugar cubes (about 2
    boxes)*
mixing bowl

mixing spoon
eggbeater
knife
sturdy cardboard base about
    12 inches by 12 inches
aluminum foil

*The project requires 385 sugar cubes. However, some cubes may break. Plan to have extra cubes available.

Procedure

1. Make a mortar by combining egg white and cream of tartar in the mixing bowl.
2. Beat with eggbeater until soft peaks form.
3. Slowly add confectioners' sugar, small amounts at a time, until a type of icing forms.
4. Cover cardboard base with aluminum foil.
5. Spread a thin layer of icing on the base.
6. Place 100 cubes on the icing in 10 rows of 10 each. Place a small amount of icing between each cube before placing it next to another cube.
7. Spread a thin layer of icing on top of the first row.
8. The next layer should be made of 81 sugar cubes in 9 rows of 9 each. Remember to place a small amount of icing between cubes.
9. The next layer should be 64 sugar cubes in 8 rows of 8 each.
10. Continue each layer (49 cubes in 7 rows of 7 each, 36 cubes in 6 rows of 6 each, 25 cubes in 5 rows of 5 each, 16 cubes in 4 rows of 4 each, 9 cubes of 3 rows of 3 each, 4 cubes in 2 row of 2 each) until the last layer has 1 cube.
11. Allow to harden before moving.

■ Rainsticks
[Makes 1]

NOTE: Rainsticks were created by Native Americans. Rainsticks can be purchased in stores in the American Southwest, but they can be very expensive. This project costs very little, and the sound is very soothing.

Materials

Paper towel tube or other long cardboard tube
small piece of cardboard or oaktag
20 small nails
scissors

masking tape
1 cup small dried beans or dried corn
markers or other items to decorate finished product

Procedure

1. Cut two cardboard circles the size of the openings of the tube.
2. Tape one of the circles to one end of the tube.
3. Insert small nails at various places along tube.
4. Pour in small dried beans or dried corn.
5. Tape other cardboard circle to the open end of the cardboard tube.
6. Decorate.
7. Turn upside down and listen to the rainstick.

■ Adobe Bricks
[Makes as many as class wants to make]

NOTE: Almost every ancient culture built structures from clay bricks dried in the sun. The southwestern Native Americans used adobe bricks to build their homes. They used adobe because that was what they had. Children could experiment and make some bricks with the straw/grass clippings and some without. Which ones are sturdier? Why?

Materials

soil, hopefully with a bit of clay in it
dry grass clippings or straw
water
bucket

1 half-gallon wax cardboard milk
   container
scissors
old knife

Procedure

1. In the bucket combine equal amounts of soil, grass clippings, and water.
2. Evenly blend the three ingredients.
3. With the scissors, cut out one of the long sides of milk carton.
4. Cut out its opposite side.
5. Force the triangular top down so that the milk carton is now rectangular. It is now an adobe mold.
6. Place the milk carton mold on a bed of grass clippings or straw.
7. Fill the milk carton mold with some of the adobe mixture and let stand an hour or two.
8. Use the knife to go around the edges of the mold and remove the carton.
9. Repeat with more adobe.
10. Do not move the bricks. Let them dry for several days.

■ Simple Loom
[Makes 1]

NOTE: Just about every culture developed a simple loom. This loom is easy to make, cheap, and durable. The warp threads remain stable, and the weft threads go under and over the warp.

Materials

1 piece very stiff cardboard about 5 inches by 7 inches	ruler
	pencil
about 24 feet of heavy duty string	scissors
yarn	transparent tape

Procedure

1. With the ruler and pencil, mark off every quarter inch along both of the 5 inch sides. These sides will become the ends of the looms.
2. Cut down along each mark about 1/2 inch to make notches.
3. Decide which side of the cardboard will be the underside of the loom. On that side tape one end of the string.
4. Wind the string around the loom and through the notches so that about 20 parallel strings now run the length of the loom. The string provides the warp for the loom.
5. Tape the end of the string on the underside of the loom.
6. Return to the top side of the loom. Cut off a piece of yarn and tie it to the extreme left string at the bottom.
7. Wind a small piece of transparent tape around the other end of the yarn, similar to the tip of a shoe lace, so that the yarn will not fray.
8. Begin weaving by going over the first warp and under the next. Repeat process until you reach the end warp thread. Then reverse directions. Continue weaving until you wish to change colors or the yarn runs out.
9. Children can make stripes or even more intricate patterns on their looms.
10. When the weaving is complete, it is time to take the completed project off the loom. Return to the underside of the loom. Cut the strings along the underside.
11. Turn the loom over to the top side. Remove 2 neighboring strings from the notches and tie them together. Continue until all the strings have been removed and knotted.
12. The strings make a nice fringe for the weaving, or they can be trimmed.
13. This weaving could make a great mug rug!

■ Iyaga (Inuit Toss and Catch Toy)
[Makes 1]

NOTE: Many cultures have a toy similar to the Iyaga. The Inuit made their game from bones and string.

Materials

1 unsharpened pencil

1 metal washer with a fairly wide hole (or rubber canning ring)

1 piece of string about 18 inches long

Procedure

1. Tie one end of the string to the pencil close to the eraser.

2. Tie the other end of the string to the washer.

3. A child holds the pencil and swings the string to try to catch the washer on the pencil.

4. A variation of this game can be achieved by adding more washers to the string.

■ Cascarones
[Makes 1]

NOTE: In Central and South America cascarones are gently cracked over people's heads during Carnival, the time just before Lent.

Materials

1 large egg (decorated if possible)

1 needle

small pair of scissors

2 tablespoons confetti

small piece of tissue paper

glue

Procedure

1. Use needle to make small hole in one end of egg.

2. Enlarge hole to about the size of a nickel.

3. Pour out egg white and yolk and wash and dry shell.

4. Pour in confetti.

5. Cover hole with a small piece of tissue paper and glue tissue paper in place.

6. Allow glue to dry.

7. Wait for Carnival and carefully crack away!

■ Paper Flowers
[Makes 6]

NOTE: Paper flowers are a part of many cultures, including those of Mexico and Central America.

Materials

18–24 pieces of tissue paper about 6 inches by 6 inches

scissors

6 pipe cleaners

Procedure

1. Cut tissue paper into irregular shapes.
2. Cut some large sizes, some medium sizes, and some small sizes.
3. Stack 3 or 4 pieces of tissue paper, large to small.
4. Make 2 small holes in the center of the tissue paper stacks.
5. Weave a pipe cleaner through the 2 holes and twist ends together to make stem.
6. Pull tissue pieces away from stem to make the flower full.
7. Complete other flowers.

■ Martenitsa
[Makes 1]

NOTE: In Bulgaria people make martenitsas to celebrate spring. People give each other bracelets made of braided red and white strings. Sometimes beads are attached. The maker of a martenitsa gives the bracelet to a good friend. People start to wear martenitsas on March 1. When someone sees either a stork or a flowering tree bud, the person removes the martenitsa and places it on the tree. Children often wear multiple martenitsas at the same time. This project is more fun when two people work together.

Materials

2 people

1 piece white string or embroidery floss about 15 inches long

1 piece red string or embroidery floss about 15 inches long

1 charm or bead with hole larger than thicknesses of flosses (optional)

Procedure

1. Tie red and white flosses together at an end, leaving about an inch of fringe at end.

2. One person holds end that has been tied. Other person holds red floss in left hand and white floss in right hand.

3. Person with untied flosses begins crossing white floss over red floss and then red floss over white floss. The flosses will begin to create a "barber pole" effect. Keep crossing flosses until effect is quite tight.

4. Insert flosses into charm or bead, if desired.

5. Tie flosses together.

6. Wrap martenitsa around someone's wrist and tie ends together. The 4 ends make nice fringe.

21
Colonial Times Projects

In 1776, about 96 percent of adults were farmers. Today about 4 percent of adults are farmers. Our society has certainly become very specialized, and daily life is very different from that of the colonial period. Hopefully these activities will help children see how colonial people lived.

■ Cornhusk Dolls
[Makes 1 doll]

NOTE: Children had to make their own toys, even their dolls. Today's children can see how easy it is to make dolls. Cornhusks can be obtained from craft catalogs, local corn farmers, or grocery stores.

Materials

12 cornhusks   scissors
container of warm water   cotton balls
thin string or embroidery floss

Procedure

1. Soak cornhusks in container of warm water for about half an hour to make them pliable.
2. Tie the 12 husks together at the top with a length of string or embroidery floss.
3. Make the head by making another tie a bit down the husks. Fill the husks with cotton balls to fill out the head.
4. Pull away 3 husks to form an arm. Tie them together.
5. Do the same for the other arm.
6. Make the body by tying together the remaining husks about halfway down.
7. Create a leg by taking 3 of those husks and tying them together just above the ends.
8. Do the same for the other leg.
9. Add cotton ball stuffing if necessary.
10. Trim the arms if necessary to make the body proportional.
11. Make clothing from scraps of cloth if desired.

■ Cornhusk Wreath
[Makes 1]

NOTE: This wreath is easy to make. The difficult part is getting the hang of knotting the cornhusks.

Materials

piece of wire about 24 inches long	masking tape
pliers	scissors
about 30 cornhusks	ribbon bow, pinecones, or other deco-
container of warm water	rations

Procedure

1. Soak cornhusks in container of warm water for about half an hour to make them pliable.
2. Shape the piece of wire into a circle. Twist together the ends with the pliers.
3. Cover the area where the ends meet with masking tape.
4. Remove cornhusks and fold each in half to make a loop.
5. Lay each folded cornhusk under the wire. Pull the tails of the husk over the wire and through the loop. Secure each knot.
6. Continue knotting on cornhusks until the wreath is full.
7. Add ribbon, pinecones, or other decorations.

■ Yarn Dolls
[Makes 1]

NOTE: Yarn dolls are soft and cuddly. They could be made much larger than the instructions indicate.

Materials

about 10 yards yarn   scissors
cardboard scrap about 8 inches wide

Procedure

1. Wind the yarn around the cardboard at least 16 times and cut at the end of the last wind.
2. Cut 7 more pieces of yarn about 10 inches each.
3. Cut yarn at bottom of the cardboard and remove cardboard.
4. Tie 1 piece of the 10-inch yarn around the yarn strands about 1 inch from the top to form the head.
5. Pull away 8 strands of yarn from the body to form an arm. Tie with a 10-inch piece of yarn to form a shoulder. Cut some of the yarn off so that the arm is not too long. Tie with a piece of the 10-inch yarn.
6. Repeat for the other arm.
7. With another piece of 10-inch yarn, tie the remaining 16 pieces to form a waist.
8. For a girl doll, trim the yarn to form the bottom of the skirt.
9. For a boy doll, pull away 8 strands at tie at the ankle with one 10-inch piece of yarn.
10. Repeat for the other leg.

■ Cup and Ball Toy
[Makes 1]

NOTE: This toy was originally made from wood. The wooden ball stung when it hit a child. Making the cup and ball is easy. Getting the ball into the cup takes coordination and patience. In this version, the ball is made out of aluminum foil so that it does not hurt the player.

Materials

1 tongue depressor   scissors
1 paper cup   hot glue gun and hot glue
1 piece of string about 18 inches long
1 piece of aluminum foil about 12
    inches by 12 inches

Procedure

1. Make a small slit in the bottom of the paper cup with the scissors.
2. Insert the tongue depressor about 1 inch into the slit and hot glue the paper cup and tongue depressor together.
3. Tie 1 end of the string around the connection of the cup and tongue depressor.
4. Crush the aluminum foil around the other end of the string and make a small ball.
5. Hold the toy by holding the tongue depressor handle. Try to swing the ball and catch it in the cup.

■ Humming Whirligig Toy
[Makes 1]

NOTE: I can make these, but I cannot make them hum. Other people take my whirligigs and make them hum nicely. I cannot skip rocks either.

Materials

flat button with at least 2 holes   scissors
30 inches of string thin enough to
    pass through button-holes.

Procedure

1. String the thread through the button holes and tie the ends. This will make a loop. The loop should be a bit longer than your body is wide.
2. Twist the thread in 1 direction until it is wound tightly.
3. Pull your arms apart and then together again. Repeat the process.
4. Soon your button will be twirling and humming.

■ Braided Rug (or Hot Pad)
[Makes 1]

NOTE: Braided rugs were a good way to cover the floor. Scraps of leftover fabric or fabric from discarded clothing were used to make brightly colored rugs.

Materials

3 long strips of fabric about 1
    inch wide

2 safety pins
needle and thread

Procedure

1. Pin together the ends of the 3 pieces of fabric with one safety pin.

2. Have another child hold the ends, or tie string to the safety pin and tie the string around something sturdy like a table leg.

3. Begin to braid the 3 strips as tightly as possible.

4. When the fabric has been completely braided, fasten the ends with the other safety pin.

5. Sew the ends together with the needle and thread.

6. Tightly curl the braid so that it forms a compact spiral. Stitch the spiral together every inch or so along the way.

■ Tin Punching
[Makes 1 project]

NOTE: Tin punching was quite an art, and it still is. Tin lanterns were punched so that the light could show through the holes. Tin plates were punched and used as decorations. This formula requires that water be frozen in the container to make it easier to punch. Children should wear safety goggles and gloves because pieces of ice could break away and fly about.

Materials

smooth can or paint container
old towel
piece of paper with pattern of dots
thin marker
water
freezer

hammer and nail
votive candle
matches
safety goggles
work gloves

Procedure

1. Fill container with water and place it in the freezer for at least a day.
2. Remove container from the freezer and place it on its side on the towel. The towel will keep it in place.
3. Children should wear safety goggles and work gloves.
4. Place the pattern on the side of the container and transfer the dots onto the container with the marker.
5. Use hammer to punch nail through the tin at every dot. Watch out for pieces of loose ice.
6. When pattern has been completely punched out, turn the container over so that the melting ice will fall from the container.
7. Allow ice to melt and empty container.
8. After container is dry, set the votive candle on bottom of inside of container.
9. Light the candle. The light should shine through the holes.

■ Quill Pen
[Makes 1]

NOTE: Writing with a quill pen is difficult. Ink sometimes goes where it should not go. It does not always go where it should. Colonial children spent about half their school day working on penmanship. Now I see why.

Materials

goose quill or any other feather at least 11 inches long (can be obtained from craft stores or poultry farms)

warm, soapy water in a small container

scissors

pin

piece of felt about 6 inches by 6 inches

ink

paper

Procedure

1. Soak bottom of quill in warm, soapy water for about 20 minutes.
2. Remove feathers from bottom 2 inches of quill.
3. Cut bottom of quill off at an angle. This will become the nib of the pen.
4. Clean inside of nib with pin.
5. Cut a slit about a half inch long at the pointed part of the nib.
6. Dip the nib into the ink and carefully blot onto the felt. This will remove excess ink.
7. To write, hold the pen at a slant. This will take practice.
8. When the nib is worn down, repeat steps to create a new sharp nib.

■ Soot Ink
[Makes about 1/4 cup]

NOTE: This ink is a simplified version of a very old recipe. This ink will not keep well unless it is refrigerated. Inks used in very old times spoiled quickly and were therefore made in small batches.

Materials

1 egg white	whisk
about 3 tablespoons soot collected from sides of fireplaces	disposable container
	disposable spoon
about 2 tablespoons honey	small jar with lid

Procedure

1. Pour egg white into disposable container. Beat egg white with whisk for just a few seconds.

2. Add soot and mix well.

3. Add enough honey to make a creamy consistency.

4. Pour into small jar and cap. Use as soon as possible or refrigerate for a few days.

■ Berry Ink
[Makes about 1/2 cup]

NOTE: This ink spoils quickly. It can be refrigerated.

Materials

1 cup ripe berries	1 teaspoon vinegar
small piece of cheesecloth	spoon
small bowl	small jar with lid
1 teaspoon salt	

Procedure

1. Line small bowl with cheesecloth.

2. Pour berries into cheesecloth.

3. Pick up cheesecloth to make a small bag and squeeze berries so that the juices trickle into the small bowl.

4. Add vinegar and salt and stir until salt dissolves.

5. Pour the berry ink into the small jar and cap. Use as soon as possible or refrigerate for a couple of days.

■ Walnut Shell Ink
[Makes about 1 cup]

NOTE: The salt and vinegar stabilize the color and keep it from spoiling.

Materials

20 walnut shells
plastic bag
hammer
2 cups water
1 1/2 teaspoons vinegar
1 1/2 teaspoons salt

old pot
stove or heating element
spoon
small piece of cheesecloth
small jar with lid

Procedure

1. Pour walnut shells into plastic bag.

2. Crush shells with the hammer.

3. Pour crushed shells into the old pot and add water.

4. Heat mixture until it boils and then let simmer for about an hour. The water should be dark brown. Allow to cool.

5. Strain mixture through the cheesecloth into the small jar.

6. Add vinegar and salt and stir to combine.

7. Cap the bottle.

■ Signet
[Makes 1 project]

NOTE: Signets were used to seal envelopes. A signet acted somewhat like a signature. Since a signet is dipped in warm wax, its impression is the opposite of what is designed.

Materials

rectangular prism of self-hardening clay about 3 inches by 1 inch by 1 inch
pencil

tweezers
vegetable oil
cotton swab
sealing wax (see next recipe)

Procedure

1. Stand clay on an end so that the top is 1 inch by 1 inch.

2. Use the pencil to lightly draw in the signet design. If a letter is used, make sure the letter is drawn in backwards.

3. Use the tweezers to dig clay away from the design so that the design stands in relief. Let the clay harden overnight.

4. Use the swab to apply a small amount of vegetable oil to the signet before stamping it into warm wax. The oil will keep the wax from sticking to the signet.

■ Sealing Wax
[Makes 1]

NOTE: Self-sealing envelopes are rather new. Years ago people made envelopes from the letters by folding the letter paper and sealing it with wax. Sealing wax was available in only 2 colors, red and black. Black sealing wax was used only when a family was mourning the death of a loved one. Red sealing wax was used for all other occasions.

Materials

handwritten letter
old candle, preferably red
match
signet (see activity above) or other

3-dimensional object to press into
hot wax, such as a coin or the tines
of a fork

Procedure

1. Fold over the corners of the letter so that they meet in the middle. The writing should be inside the folds.

2. Light old candle with match. Wait a minute or so and drip some hot wax onto the middle of the folds.

3. Quickly press the 3-dimensional object into the hot wax and remove. An impression of the object should remain.

4. Let wax cool, and letter is ready to "send."

CHAPTER
22
Soaps

People have been making soap for thousands of years. The American colonists made soap by combining rendered animal fat with lye. Making soap from such ingredients can be dangerous. Lye, leached from wood ash, is caustic, and rendering animal fat is not fun!

Therefore, making soap in traditional ways is not recommended for children. The following recipes and formulas are adaptations of true soap-making methods and will help children understand the process.

Soaps made by the following methods should be used as demonstrations and not as actual cleaning agents. Some skin types may react to these soaps in adverse ways.

■ New Soap from Old Soap

[Makes 4 soap cakes]

NOTE: In this formula, children recycle soap scraps to see how soap is made.

Materials

1 cup soap scraps cut into small
   pieces
1 cup water
stove or heating element
steel pan

mixing spoon
essential oils (optional)
small, disposable aluminum loaf or
   pie pans

Procedure

1. Combine soap pieces and water in steel pan.

2. Heat mixture until it boils.

3. Add essential oils if you want your soap to have a pleasant smell.

4. Let mixture simmer, stirring constantly, for 5 minutes. The mixture will take the shape of a ball.

5. Remove from heat and let cool for 10 minutes.

6. Pour into disposable pans to make the soap into cakes. Squeeze out bubbles.

7. Allow soap cakes to age for about a month. Soap cakes should easily slide from pans.

■ Powdered Soap to Hard Soap 1 (Laundry Soap and Vegetable Oil)
[Makes 2 cups]

NOTE: Powdered laundry soap becomes hard soap in about 15 minutes.

Materials

1 1/2 cups vegetable oil
1 cup powdered laundry soap (not
   laundry detergent)
stove or heating element

steel pot with lid
mixing spoon
wax paper

Procedure

1. Stir soap and vegetable oil together in steel pot.

2. Simmer for about 15 minutes with lid on pot. Do not take lid off until simmering is complete.

3. Remove pot from heat and let cool.

4. When pot is cool, remove soap. Shape soap and let harden on wax paper for several days.

■ Powdered Soap to Hard Soap 2 (Laundry Soap and Salt)
[Makes 1 cup]

NOTE: This formula is easier than the one above. The dissolved soap becomes solid soap very quickly.

Materials

5 tablespoons laundry soap (not
   laundry detergent)
2 cups hot water
1/4 cup salt

2 mixing bowls
mixing spoon
wax paper

Procedure

1. In one bowl, dissolve laundry soap in 1 cup of hot water.

2. In other bowl, dissolve salt in 1 cup hot water.

3. Pour salt mixture into soap mixture. Do not stir.

4. After a while, solid soap will form on top of mixture.

5. Remove solid soap from bowl and let dry on wax paper. This drying could take several days.

■ Liquid Soap to Hard Soap

[Makes 1/2 cup]

NOTE: The liquid soap will become hard soap very quickly. Children are often amazed to see this process.

Materials

3 tablespoons salt

1 cup hot water

3 tablespoons liquid castile soap
  (found in health food stores or soap
  specialty shops)

mixing bowl

mixing spoon

cheesecloth

Procedure

1. Combine salt and hot water in mixing bowl.

2. Pour in soap and stir.

3. Within minutes, hard soap will float on top.

4. Remove hard soap and place in piece of cheesecloth.

5. Hang cheesecloth so soap will dry. This may take several days.

■ Powdered Soap to Liquid Soap

[Makes 3 cups]

NOTE: The liquid soap has a slippery consistency

Materials

2 cups laundry soap (not laundry
  detergent)

2 cups hot water

2 tablespoons baby oil

essential oils (optional)

mixing bowl

mixing spoon

liquid soap dispenser

Procedure

1. Pour laundry soap into bowl.

2. Add water and mix. Stir in baby oil.

3. Add essential oils if desired.

4. Pour into liquid soap dispenser.

5. Shake once in a while to keep materials from separating.

■ Soap Balls
[Makes 4 balls]

NOTE: Making soap balls is fun and simple. For decorative and aromatic purposes only, these balls make wonderful presents for Mother's Day or Valentine's Day. Children could wrap the finished soap balls in netting and tie with ribbons.

Materials

2 cups laundry soap (not laundry detergent)
food coloring (optional)
essential oils (optional)

2 tablespoons water
mixing bowl
wax paper

Procedure

1. Pour soap into bowl. Add food coloring, essential oils, and water.
2. Mix with hands until a ball is formed.
3. Divide big ball into 4 balls.
4. Allow to dry for several days on wax paper.

■ Glycerin Soap

[Makes 2 bars]

NOTE: Glycerin soap base can be purchased at craft stores. You can make your own base, but the prepared base is much, much easier. You can buy attractive soap molds, but you can use anything plastic as a mold. Examples include ice cube trays or clean yogurt containers.

Materials

8 1-inch by 1-inch blocks clear
  glycerin soap base
food coloring (optional)
essential oils (optional)

microwavable container
mixing spoon
2 soap molds
microwave oven

Procedure

1. Place glycerin soap base blocks in microwavable container. Microwave at high for 40 seconds and then at 10 second intervals until cubes have melted.

2. Remove container from microwave and add food coloring (optional) or essential oils for fragrance (optional). Remember that a few drops of essential oil go a long way.

3. Pour melted base into molds.

4. Let harden for 40 minutes.

5. Pop hardened soaps from molds and let age for a week or so.

CHAPTER

23
Candles

241

Many families in colonial America made candles from tallow, rendered animal fat. Today, however, most candles are made from beeswax or paraffin. Beeswax candles last longer and give a brighter light than paraffin candles, and they have a lovely fragrance. However, as might be expected, beeswax costs more than paraffin. To improve paraffin candles, add some beeswax to the mixture. Just 20 percent beeswax in the mixture makes a difference in the end product. If using paraffin alone, add 3 tablespoons stearic acid (which comes in a powder form) to each pound of paraffin. The stearic acid reduces smoke and makes candles harder. Both beeswax and paraffin are available from craft stores and candle-making supply houses. Paraffin can also be purchased in grocery stores, hardware stores, and chemical supply houses.

Two different types of wicks, wire and braid, are available in various thicknesses. Use thinner wicks for short or thin candles and thicker wicks for tall or wide candles. Craft stores sell many supplies to facilitate candle making. Items include wick holder tabs, concentrated colors, scents, mold sealers, and mold releases.

■ Wax Snow

NOTE: Because more wax is being applied to an existing candle, wax snow is an excellent beginning candle project. The finished candles brighten a holiday setting.

Materials

beeswax or paraffin	electric frying pan
water	hot pads
3 tablespoons stearic acid per pound of wax	eggbeater
	spoon
coffee can	candle with a fairly wide diameter

Procedure

1. Put wax in coffee can.
2. Pour several inches of water into frying pan. Place coffee can in frying pan so that the two act as a double boiler. Slowly melt wax. Watch out that wax does not sputter or burn someone.
3. When wax has melted, turn off heat. Add stearic acid.
4. Using hot pads, remove coffee can of wax.
5. Allow to cool until a film forms on surface. Beat with eggbeater until wax looks like snow.
6. Lift snow wax with spoon onto large candle. Wax snow can be applied to either the entire candle or just parts of the candle. Do not cover wick. Let wax harden.

■ Floating Candles

NOTE: These candles make a lovely centerpiece for a party table. They also make an interesting science experiment. Pour water into a large heat-resistant bowl (e.g., punch bowl). Place these small, lightweight candles on the surface of the water. Light the candles, and observe the effects.

Materials

wax snow (from previous formula)   wick holder tabs (from craft store)
muffin tins   spoon
wick

Procedure

1. Make wax snow as described above.
2. Cut wick into pieces 2 inches longer than muffin tin cups are deep.
3. Attach wicks to wick holder tabs and fasten 1 to inside bottom of each muffin tin cup.
4. Spoon wax snow into each muffin tin cup. Keep wicks vertical.
5. Allow candles to dry and remove from muffin tin.

■ Dipped Candles

NOTE: Dipped candles are easy to make with large groups. Organize children into groups of three or four. Each child gets one wick, and each group receives one stick. Each child ties his/her wick to the group's stick. Place groups in a large circle. As one group dips, the other groups are in different stages of waiting for their candles to harden or waiting to dip again.

Materials

beeswax or paraffin

3 tablespoons stearic acid per pound of wax

scent (optional; available in blocks at craft stores)

color (optional; use crayons or old candle stubs, or purchase in blocks at craft stores)

wicks, 3 inches longer than the desired candles' length

water

large coffee can, deeper than the desired candles' length

electric frying pan

candy thermometer

wooden mixing spoon

wooden dowels about 24 inches long

bucket full of room temperature water

refrigerator

old knife

Procedure

1. Put wax in coffee can.

2. Pour several inches of water into frying pan.

3. Place coffee can in frying pan so that the two act as a double boiler. Slowly melt wax, keeping wax temperature lower than 180°F. Watch out that wax does not sputter or burn someone.

4. When wax has melted, turn off heat. Add stearic acid. Add scent or color if desired. Stir thoroughly with wooden mixing spoon.

5. Cut 3 pieces of wick and tie to wooden dowel, letting wicks hang down.

6. For first coat, dip wicks in wax, let wicks soak in wax for 30 seconds and remove. Let wax harden. Dip wicks into bucket of room temperature water between wax dippings, or refrigerate candles between dippings.

7. Dip wicks again into wax. A new layer will be added. Let this wax harden. If wax is too hot, it will melt previous layer. If wax becomes too hard, slightly heat it again.

8. Keep dipping and hanging (about 15 times) until desired candle diameter is reached.

9. Let candles harden.

10. Cut bottom of each candle to form flat surface.

11. Snip wick knots to free candles from stick. Candles are done.

■ Molded Candles

NOTE: Craft stores sell two-part candle molds with special releases and wonderful designs. However, many items at home and school make terrific molds. Examples are plastic tubs, milk cartons, and smooth metal containers.

Materials

vegetable oil
beeswax or paraffin
3 tablespoons stearic acid per pound of wax
scent (optional; available in blocks at craft stores)
color (optional; use crayons or old candle stubs, or purchase in blocks at craft stores)
wicks, 3 inches longer than height of mold
candle mold

craft stick, metal rod, or knitting needle longer than width of mold
washer (from hardware store)
clay
metal pitcher or coffee can with top edge pinched to form a spout
electric frying pan
water
hot pads
refrigerator
old, soft towel

Procedure

1. Wipe inside of mold with vegetable oil so that candle will not stick to mold.
2. Make a hole large enough to accommodate wick in bottom of mold.
3. Insert wick through hole.
4. On outside of mold, tie washer to end of wick.
5. Fill hole with clay so that liquid will not seep through hole.
6. Turn mold over and stand it on its bottom. Tie top of end of wick on craft stick, metal rod, or knitting needle. Lay craft stick, metal rod, or knitting needle across width of mold. Make sure wick is taut and perpendicular to bottom of mold.
7. Put wax in coffee can.
8. Pour several inches of water into frying pan.
9. Place coffee can in frying pan so that the two act as a double boiler. Slowly melt wax. Watch out that wax does not sputter or burn someone.
10. When wax has melted, turn off heat. Add stearic acid. Add scent or color if desired. Stir thoroughly with wooden mixing spoon.
11. Using hot pads, pour wax into mold, but do not disturb wick.
12. Let wax cool for 24 hours.
13. A slight depression may occur where wick meets candle. Melt a bit of wax again and pour around wick.
14. Let candle cool again for 24 hours.

15. Put candle in refrigerator for 24 hours.

16. Slide candle out of mold. If mold does not give away easily, dip mold into hot water.

17. Smooth down any rough edges of candle by rubbing spots with an old, soft towel.

18. Candles should age for at least a week.

■ Shaped Candles

NOTE: These candles are great fun because children can shape the warm wax to suit their fancies. Carefully monitor temperature of the wax.

Materials

beeswax or paraffin	wicks
3 tablespoons stearic acid per pound of wax	metal pitcher or coffee can with top edge pinched to form a spout
scent (optional; available in blocks at craft stores)	electric frying pan
	water
color (optional; use crayons or old candle stubs, or purchase in blocks at craft stores)	hot pads
	carving tools
	knife

Procedure

1. Put wax in coffee can.

2. Pour several inches of water into frying pan.

3. Place coffee can in frying pan so that the two act as a double boiler. Slowly melt wax. Watch out that wax does not sputter or burn someone.

4. When wax has melted, turn off heat. Add stearic acid. Add scent or color if desired. Stir thoroughly with wooden mixing spoon.

5. Using hot pads, remove can of wax from electric frying pan and cool on table or counter.

6. Let wax cool for about 40 minutes, occasionally puncturing surface to allow wax beneath to cool faster.

7. Test wax to see if it is cool enough to handle. Give each child a portion of wax.

8. Have children manipulate warm wax by sculpting and carving wax around wick.

9. Let candles stand for a week before using.

10. Trim off bottom of each candle with knife to make a flat surface on candle.

■ Sand-Cast Candles

NOTE: A warm, sunny day on a playground of sand is a good place to make sand-cast candles. These candles can also be made indoors, using buckets or small containers of sand.

Materials

sand wet enough to make sand castles (out in the yard or in containers)

wick at least 4 inches longer than desired length of candle

weight such as a washer or nut

craft stick, metal rod, or knitting needle longer than width of hole in the sand

beeswax or paraffin

3 tablespoons stearic acid per pound of wax

scent (optional; available in blocks at craft stores)

color (optional; use crayons or old candle stubs, or purchase in blocks at craft stores)

wicks

metal pitcher or coffee can with top edge pinched to form a spout

electric frying pan

water

hot pads

Procedure

1. Using this method, the bottom of the hole in the sand will be the top of the completed candle.

2. Tie one end of wick to weight.

3. Make a depression in sand. Poke holes with fingers. When satisfied with shape of sand mold, bury weight and end of wick. Make sure wick goes 1–2 inches into sand. Firmly pack sand; otherwise, wax just seeps into sand rather than remaining in hole.

4. Tie other end of wick to craft stick, metal rod, or knitting needle. Lay craft stick, metal rod, or knitting needle across width of hole in sand. Make sure wick is taut.

5. Put wax in coffee can.

6. Pour several inches of water into frying pan.

7. Place coffee can in frying pan so that the two act as a double boiler. Slowly melt wax. Watch out that wax does not sputter or burn someone.

8. When wax has melted, turn off heat. Add stearic acid. Add scent or color if desired. Stir thoroughly with wooden mixing spoon.

9. Using hot pads, remove can of wax from electric frying pan and pour wax into hole. Try not to disturb wick.

10. Let wax cool for at least a day.

11. Slowly remove sand away from candle. Some sand should cling to candle.

12. Cut weight on wick.

13. Turn candle over. Make sure candle bottom is flat.

■ Beeswax Sheet Candles

NOTE: Children enjoy making these candles. The candles require no heat, and a group of children can make them at the same time. The candles burn for a long time and produce little smoke. Sheets of beeswax are about 15 inches long, but they can be cut into shorter lengths.

Materials

sheets of beeswax   wicks

Procedure

1. The beeswax sheets should be pliable. Usually body heat from fingers and hands is enough to soften the sheets. However, the sheets can be slightly warmed in an oven or microwave if necessary.

2. Lay a sheet of beeswax flat on a table. Cut one side of the sheet so that there is a slight slant of 1 to 2 inches from one side to the other.

3. Place the wick atop the longer side of the sheet.

4. Roll longer edge of wax sheet over wick and keep rolling. A tube of wax forms with the wick in the center. The sides will slant down from the wick.

5. This candle does not have to age.

■ Rolled Paraffin Candles

[Makes about 4]

NOTE: These candles are easy to make, and they can be quite colorful.

Materials

beeswax	hot pads
old crayons (with paper removed) or candle stubs	knife
	small cookie sheet with lip
metal pitcher or coffee can with top edge pinched to form a spout	wax paper
	wick
electric frying pan	small washers
water	

Procedure

1. Put beeswax in coffee can.

2. Pour several inches of water into frying pan.

3. Place coffee can in frying pan so that the two act as a double boiler. Slowly melt wax. Watch out that wax does not sputter or burn someone.

4. When wax has melted, turn off heat. Stir thoroughly with wooden mixing spoon.

5. If desired, add old crayons or candle stubs for color.

6. Tie wick to the small washers. The washers will weigh down the wick.

7. Line cookie sheet with wax paper.

8. Use hot pads to pour melted beeswax onto the cookie sheet. Consider pouring 2 colors of wax next to each other and marbleizing the material with a tongue depressor.

9. Let wax cool but not harden.

10. While wax is still warm and pliable, cut it into sections. Place a wick with washer along one edge of a section of wax.

11. Roll wax over the wick so that ultimately the wick is in center of candle.

12. Stand candle up and let cool.

■ Egg-Shaped Candles

NOTE: Egg shells provide molds for these candles. Easy to make, these candles are especially nice for Easter activities.

Materials

raw eggs
wick
scissors
small piece of clay
egg carton
pencil
beeswax or paraffin
3 tablespoons stearic acid per pound of wax
scent (optional; available in blocks at craft stores)

color (optional; use crayons or old candle stubs, or purchase in blocks at craft stores)
metal pitcher or coffee can with top edge pinched to form a spout
electric frying pan
water
hot pads

Procedure

1. Carefully tap off pointed end of raw eggs. Pour out yolks and egg whites.

2. Wash and air dry egg shells.

3. Cut a piece of wick about 2 inches longer than the depth of the egg shell. Fasten wick to inside bottom of shell with a bit of clay. Place egg shell upright in egg carton.

4. Tie top of wick to pencil and lay pencil across top of egg shell.

5. Repeat process for all the eggs.

6. Pour several inches of water into frying pan.

7. Place coffee can in frying pan so that the two act as a double boiler. Slowly melt wax. Watch out that wax does not sputter or burn someone.

8. When wax has melted, turn off heat. Add stearic acid. Add scent or color if desired. Stir thoroughly with wooden mixing spoon.

9. Using hot pads, remove can of wax from electric frying pan and pour wax into egg shells. Try not to disturb wicks.

10. Let wax cool for at least a day. Peel off shells, or leave shells in place as natural containers.

■ Salt Candles
[Makes 1]

NOTE: The heat-resistant jar can become hot when the candle is lit.

Materials

clean, empty heat-resistant jar
3/4 cup salt
3 colors of food coloring

3 small empty yogurt containers
mixing spoons
1 votive candle

Procedure

1. Distribute salt among yogurt containers and color with food coloring.
2. Pour each layer of salt carefully into jar. Do not mix layers.
3. Push votive candle into layers of salt.

■ Seashell Candles
[Each shell makes 1 candle]

NOTE: These candles make good reminders of summer vacations. They are also a way to recycle nature's gifts.

Materials

several layers of newspaper
several clean seashells
mixing spoons
beeswax
empty can in which to melt wax

double boiler
water for double boiler
heating element
hot pads
birthday candles

Procedure

1. Spread several layers of newspaper out on the working surface.
2. Place seashells on the newspaper. Prop up any shells that might tip over with wads of newspaper.
3. Place beeswax in empty can and place it in the double boiler. Add water to the double boiler. Melt wax over low heat.
4. Use hot pads to pour wax into the seashells.
5. Let wax cool for about 5 minutes.
6. Place a birthday candle into the wax. This will form the wick of the candle. Do not worry if the top of the birthday candle shows. It will melt the first time the candle is lit.
7. Let candles cool.

■ Squash Candles
[Makes 1]

NOTE: These candles are great for harvest activities.

Materials

1 squash or gourd
knife
several layers of newspaper
beeswax
empty can in which to melt wax
double boiler

water for double boiler
heating element
hot pads
wick
small washer

Procedure

1. Spread several layers of newspaper out on the working surface. Cut off top of squash or gourd. Remove all possible flesh.

2. Place squash or gourd on newspaper. Prop up any sides that might tip over with wads of newspaper.

3. Place beeswax in empty can and place it in the double boiler. Add water to double boiler. Melt wax over low heat.

4. Tie wick to small washer. The washer will weigh down wick in the hot, liquid wax.

5. Place washer on the bottom of the squash. Make sure the free end of the wick rests on the edge of the squash.

6. Pour wax into the squash.

7. Let wax cool. Check every few minutes for an hour or so to make sure wick is vertical.

24

Make Your Own Paper and Paper Projects

Paper originated about 2,000 years ago. The Chinese were the first to discover how to make paper. Before that, papyrus, vellum, parchment, bark, and various other materials were used as writing surfaces. Paper is the result of beating plant fibers, bringing forth the cellulose in those fibers, adding water, and sieving the mixture, which is called slurry.

Making new paper from old paper and other materials is fun, but it is also messy and it takes practice. First products are generally quite thick and resemble the material from which egg cartons are made.

Mold and Deckle

The mold, basically a sieve, strains paper fibers from slurry. The deckle, which snaps onto the mold, frames the new paper and keeps the paper fibers on the mold as it is lifted from the slurry. Mold and deckle sets are available at craft stores and through art supply catalogs. If you do not wish to purchase a mold and deckle set, you can easily make your own.

■ Traditional Rectangular Mold and Deckle
[Makes 1]

NOTE: This process takes the most work of options given. However, it is the most durable.

Materials

window screening, 12 inches by 14 inches

wood molding

nails

staple gun and staples

Procedure

1. Make two frames from wood molding and nails. Make one frame 10 inches by 12 inches. Make other frame slightly smaller.

2. Wrap window screening around larger of 2 frames.

3. Staple window screening onto frame to make the mold.

4. The other frame (without screening) is the deckle.

■ Easy Rectangular Mold and Deckle
[Makes 1]

NOTE: Make sure picture frames are free of glass and backing.

Materials

2 picture frames (one frame should be slightly larger than the other)
scissors

1 piece plastic canvas (7-mesh; available at craft stores)
hot glue gun and glue

Procedure

1. Cut plastic canvas to fit on top of larger picture frame.
2. Hot glue plastic canvas to frame. This becomes mold.
3. The other empty picture frame becomes deckle.

■ Easy Round Mold
[Makes 1]

NOTE: Check craft stores for embroidery hoops.

Materials

pair of embroidery hoops
netting, slightly larger than the larger embroidery hoop

Procedure

1. Separate embroidery hoops.
2. Lay netting over smaller, inner embroidery hoop.
3. Place larger hoop over netting and snap hoops and netting together.

■ Very Easy Round Mold and Deckle
[Makes 1]

NOTE: The handles of the grease-spatter screens make paper-making tidier.

Materials

2 grease-spatter screens
(available in kitchenware
stores)

craft knife

Procedure

1. Keep one spatter screen just as it is. It will be the frame.

2. Using craft knife, cut away screen from second spatter screen. It will be the deckle.

Paper and Paper Products

■ Basic Recycled Paper

NOTE: Plan 2 days for this project. Children love the mess, and they are amazed they can actually make paper.

Materials

newspaper
warm water
bucket
scoop
blender

mold and deckle
old dishpan wider than mold and deckle
damp cloth

Procedure

1. Tear newspaper into pieces the size of quarters. Place pieces into bucket.
2. Add warm water to cover paper. Soak paper overnight in bucket.
3. The next day, scoop 1 cup of paper into blender. Cover with water. Blend until pulpy. This combination of pulp and water is called slurry.
4. Add more paper and blend again.
5. Pour slurry into dishpan.
6. Repeat process until dishpan is half full.
7. Slide mold and deckle through slurry until screen rests on bottom of dishpan.
8. With both hands, raise mold and deckle. A great deal of slurry should stay on mold as it is raised.
9. Press slurry to get rid of water and to distribute slurry evenly.
10. When desired shape and thickness are attained, remove deckle. Flip mold over damp cloth. The new paper will fall onto damp cloth.
11. Repeat process to make more sheets of paper.
12. Allow paper to dry. Remove from cloth.

■ Bleached Paper

NOTE: Basic recycled paper is rather gray. In this formula, the bleach makes the paper whiter.

Materials

materials from "Basic Recycled
Paper" (See p. 257)

small amount of bleach*
rubber gloves

*Bleach should be handled very carefully. It can damage fabrics, and it can harm skin.

Procedure

1. Follow directions for "Basic Recycled Paper." However, add a small amount of bleach to paper before it soaks overnight. Children should wear rubber gloves if they are near the bleach.

■ Rag Paper

NOTE: Rag paper, which contains cotton fibers, is stronger than paper made exclusively from traditional plant material. Collect lint from a clothes-dryer filter.

Materials

materials from "Basic Recycled
Paper" (see p. 257)

approximately 1 cup clothes-dryer
lint for every 1/2 gallon slurry

Procedure

1. Follow directions for "Basic Recycled Paper." However, add clothes-dryer lint to paper before it soaks overnight.

■ Sized Paper

NOTE: Purchased paper has been sized, meaning ink applied to the paper will become part of the paper. Ink applied to unsized paper can run or bleed. This paper must be sized to accept the ink. The process of sizing is easy, but the amount of sizing added to the slurry may have to be adjusted. Sizing can be completed internally (in the beginning pulp mixture) or externally (after sheets of paper have been formed). This formula uses internal sizing.

Materials

1 envelope (1 tablespoon) unflavored gelatin
1 cup hot water
small mixing bowl
mixing spoon
materials from "Basic Recycled Paper" (see p. 257)

Procedure

1. Combine gelatin and hot water in small bowl.

2. Follow directions for "Basic Recycled Paper." However, add gelatin–water mixture to slurry before inserting mold and deckle.

■ Multicolored Paper

NOTE: This recycled paper makes unusual wrapping paper.

Materials

materials from "Basic Recycled Paper," except newspaper (see p. 257)
brightly colored paper from catalogs, magazines, and newspaper inserts
several tablespoons white glue

Procedure

1. Follow directions for "Basic Recycled Paper." However, substitute small pieces of torn paper from catalogs, magazines, and newspaper inserts for the newspaper.

2. Add white glue to slurry before inserting mold and deckle.

■ Colored Paper

NOTE: Because natural dyes are used in this formula, the color may bleed if the recycled paper becomes wet.

Materials

materials from "Basic Recycled Paper" (see p. 257)

natural dye (see chapter 8)

Procedure

1. Follow directions for "Basic Recycled Paper." However, add natural dye to paper before it soaks overnight.

■ Paper from Plants

NOTE: In this formula the fibrous plant material bonds with the pulp and becomes part of the body of the paper.

Materials

several cups of a fibrous plant material, such as onion skins, dried corn husks, celery stalks, day lily stems, iris leaves, or pineapple leaves
water
knife or scissors

materials from "Basic Recycled Paper" (see p. 257)
pot
mixing spoon
sieve
stove or heating element

Procedure

1. Cut plant fibers into pieces no longer than 2 inches. Place pieces in pot.
2. Cover fibers with water and boil until soft. This process takes about an hour.
3. Using sieve, strain out plant fibers.
4. Follow directions for "Basic Recycled Paper." Add some of the plant fibers in the blender before adding soaked paper.

■ Paper with Accents from Nature

NOTE: The difference between this formula and the formula above is that pine needles, etc. add visual interest to the paper. In the formula above, the plant fibers become a part of the paper's structure. The natural materials often add an interesting scent to the paper. Children enjoy finding the natural materials in the final product.

Materials

several cups of small, flat natural items, such as pine needles, flower petals, leaves, herbs, ferns

materials from "Basic Recycled Paper" (see p. 257)

Procedure

1. Follow directions for "Basic Recycled Paper." However, add natural materials to slurry before inserting mold and deckle.

■ Potato Paper
[Makes enough for about 8 pieces of paper]

NOTE: Potatoes have a great deal of starch in them. The potato starch helps bind the paper fibers together.

Materials

6 large raw potatoes
peeler
grater

materials from "Basic Recycled Paper" (see p. 257)

Procedure

1. Peel and grate raw potatoes.

2. Follow directions for "Basic Recycled Paper." However, combine equal amounts of slurry and grated raw potatoes. Continue to follow directions for "Basic Recycled Paper."

■ Watermarked Paper

NOTE: Watermarks have nothing to do with water. A watermark is produced by adding a shape to the mold. When the mold is pulled out of the slurry, less fiber is above the shape than the rest of the mold. Therefore, the section on the shape is more translucent than the rest of the paper.

Materials

materials from "Basic Recycled Paper" (see p. 257)

small, flat items such as a washer
small pieces of wire

Procedure

1. Attach washer to mold with small pieces of wire.
2. Follow directions for "Basic Recycled Paper." Look for the watermark when paper is dry.

■ Embossed Paper

NOTE: Embossed papers have a raised or textured surface. Embossing may be as simple as an initial or as complicated as a mask.

Materials

materials from "Basic Recycled Paper" (see p. 257)
second piece of cloth

3-dimensional form, such as a raised metal letter

Procedure

1. Follow directions for "Basic Recycled Paper."
2. Place paper on damp cloth and let dry for several hours.
3. Place 3-dimensional form onto second cloth.
4. Place still-damp paper over 3-dimensional form. Mold paper over form to make sure paper fibers conform to shape.
5. Let paper dry thoroughly before removing from form.

■ Decorative Paper Containers

NOTE: Recycled paper is somewhat similar to papier-mâché in that it can be shaped over containers. When dry, it takes the shape of the container. The finished product is not waterproof.

Materials

materials from "Basic Recycled Paper" (see p. 257)

scoop

sieve

small bowl, plate, or other container to serve as mold

scissors

paints and paintbrushes (optional)

Procedure

1. Follow directions for "Basic Recycled Paper." However, do not use mold and deckle.
2. Remove a scoopful of slurry and drain water by using sieve.
3. Apply slurry to small bowl, plate, or other container and press it around the container either on the inside or the outside.
4. If the slurry is molded around the inside of the container, the finished product will be rough on the inside and smooth on the outside.
5. If the slurry is molded around the outside of the container, the finished product will be rough on the outside and smooth on the inside.
6. Let stand 1 day, Remove original container and trim edges of finished product with scissors.
7. Paint if desired.

■ Deckled Paper

NOTE: One way to age paper is to deckle the edges. The paper's edges will look frayed, typical of old paper.

Materials

1 sheet paper	sponge
water	ruler

Procedure

1. Slightly dampen edges of paper with sponge.
2. Place ruler on the paper, parallel to and near the paper's edge.
3. Hold the ruler down firmly with one hand and tear the paper against the ruler with the other hand. The paper will have a ragged edge.
4. Let paper dry.

■ Aged Paper

NOTE: This process makes new paper look old. Children can use the technique to age newly created letters, maps, or diary entries.

Materials

2 cups fairly hot coffee or tea	large bowl
written project such as a letter, map, or diary entry	newspaper

Procedure

1. Pour coffee or tea into bowl.
2. Crumple written project and submerge in coffee or tea.
3. Let stand for a short time. The longer the soak, the darker the tint. However, the paper also becomes more fragile.
4. Remove paper from liquid and dry on newspaper.

■ Marbled Paper

NOTE: Marbled paper is easy to make—simply swirl some paints on top of a liquid base. Then carefully place a piece of plain paper over it. Children like to see the "what if" possibilities of creating patterns and whirls and checking them out.

What can children do with the marble paper they make? They can cover homemade books. They can wrap the paper around clean, empty cans and make pencil holders. Bookmarks, wrapping paper, and paper jewelry are all possibilities.

Never use detergent to clean the supplies. Soap of any kind can upset the next marbling session.

Materials

newspaper
pan, 9 inches by 12 inches
liquid starch
food coloring

tools such as combs, toothpicks, or
  paint brushes to make patterns
blank paper
sink with running water

Procedure

1. Spread newspaper over work area.
2. Pour liquid starch into pan so that starch is at least 1 inch deep. The liquid starch is called the size when marbling.
3. Dot food coloring onto surface of liquid starch.
4. Using tools, comb through the size and food coloring to make swirls and other patterns.
5. Carefully place blank paper on the surface of the marbled size.
6. Let paper rest for 30 seconds. Carefully lift paper off.
7. Turn paper over so that the colored side is face up.
8. Let coloring soak into paper for 1 minute. Rinse with tap water to remove excess coloring/starch.
9. Dry and use.

■ Shades of Gray Marbled Paper

NOTE: This process gives a range of shades of gray and black. The water base makes it extremely easy to use.

Never use detergent to clean the supplies. Soap of any kind can upset the next marbling session.

Materials

newspaper
pan, 9 inches by 12 inches
water
India ink

tools such as combs, toothpicks, or
    paint brushes to make patterns
blank paper
sink with running water

Procedure

1. Spread newspaper over work area.
2. Pour water into pan so that water is at least 1 inch deep.
3. Dot India ink onto surface of water.
4. Using tools, comb through the size and food coloring to make swirls and other patterns.
5. Carefully place blank paper on the surface of the marbled solution.
6. Let paper rest for 30 seconds. Carefully lift paper off.
7. Turn paper over so that the colored side is face up.
8. Let India ink soak into paper for 1 minute. Rinse with tap water to remove excess India ink.
9. Dry and use.

Paper Projects

■ Pinwheel
[Makes 1]

NOTE: This project has been around for a very long time. A windy spring morning would be great weather for pinwheels. Children can decorate the paper before it is bent and fastened.

Materials

1 piece square construction paper
    about 10 inches on each side
scissors

1 straight pin
1 pencil with a soft eraser

Procedure

1. Fold paper on one diagonal to make a triangle. Press to make a crease.
2. Open paper and fold on other diagonal. Press to make a crease.
3. Open paper again.
4. The center is where the 2 folds meet.
5. Cut along the folds close to but not through the center. You should have 4 triangles joined at the center.
6. Take one of corner cuts and bring it to the center. Hold it in place.
7. Do the same for the corresponding corners of the other 3 triangles.
8. Push straight pin through corners, through the center, and into the pencil eraser.
9. Make sure the straight pin allows the pinwheel to turn freely.

■ Vertical Spinners
[Makes 1]

NOTE: This project, amazingly simple to set up, produces such excitement. This experiment lends itself to many further investigations. Can we add more paper clips? Can we shorten the wings? Can we make it from other materials? Children can devote quite a few science sessions to setting up and conducting experiments.

Materials

1 piece stiff paper about 2 inches by 6 inches

medium size paper clip
scissors

Procedure

1. Hold the paper on 1 long side and cut down one short end about 2 1/5 inches.

2. About an inch below that cut, snip the two long sides about 3/4 inch each.

3. Go back to the first cut and fold one flap one direction and the other flap the other direction. These form the "wings" of the twirler.

4. Now go to the second cut. From one cut fold the paper lengthwise in the direction opposite that of its corresponding wing.

5. Fold the paper lengthwise from the other small cut in the direction opposite that of its corresponding wing.

5. Fold up bottom inch of twirler and add a paper clip.

7. Hold twirler as high as possible just under wings and drop it. It spins as it drops.

■ Horizontal Spinners
[Makes 1]

NOTE: This project also causes great excitement. Can we make it smaller? How big can we make it? Does it work without paper clips?

Materials

1 strip of construction paper, 1 inch   1 paper clip
   by 7 inches

Procedure

1. Fold paper in half so that it is 1 inch by 3 1/2 inches.
2. Approximately 1 inch from the fold end fold down the end at a 45 degree angle.
3. Approximately 1 inch from the unfolded end fold down the end at a 45 degree angle but in the opposite direction from the other end.
4. Fasten paper clip in middle of paper.
5. Open up the two ends slightly to form wings.
6. Drop paper and watch it spin.

■ Circular Spinners
[Makes 1]

NOTE: These spinners can become quite colorful if crayons or markers are used to decorate the circle.

Materials

manila folder or piece of somewhat   compass
   stiff cardboard   scissors

Procedure

1. Use compass to create an 8-inch circle on the manila folder.
2. Cut out circle.
3. Cut 6 evenly spaced slits, each about 2 1/2 inches, from the edge of the circle toward the center.
4. Now fold the circle along the slits to create flaps. Make sure the flaps all go in the same direction.
5. Hold spinner high up but horizontal to the floor and drop it.
6. It should spin as it falls to the ground.

■ Chinese Cobweb
[Makes 1]

NOTE: Make very close cuts to create an intricate cobweb.

Materials

1 large circular piece of construction   scissors
    paper or crepe paper

Procedure

1. Fold construction or crepe paper circle in half and in half again. Fold for a third time.
2. Alternate cutting slits from each length. Each cut should reach within an inch of the other side.
3. Unfold the paper and pull to make the cobweb.

■ Flower Cages
[Makes 1]

NOTE: These cobweb cutouts are called flower cages because pictures of flowers were often hidden below the cut circles of paper.

Materials

construction paper   paper clip
compass   picture of flowers
scissors   glue
string

Procedure

1. Use compass to make a large circle of construction paper. Cut it out.
2. Fold circle in half and then in half 2 more times.
3. Use scissors to make a curved cut from 1 fold almost to the other side.
4. Make another cut from the opposite side almost to the other fold.
5. Continue making alternate, curved cuts until the point is reached.
6. Make a small hole at the point of the folded paper. Unfold the paper.
7. Tie the paper clip to the string and gently place the paper clip through the small hole.
8. Glue the edges of the circle onto a picture of flowers. Glue that onto another piece of construction paper.
9. Pull and watch the paper form a cage. The flowers should appear to peek out from the cage.

■ Flower Cage Garlands
[Makes 1]

NOTE: These flower cage garlands would be great to make during Chinese New Year.

Materials

paper circles cut in flower cage method (procedure above)	glue string

Procedure

1. Place 1 flower cage circle on table. Glue at 4 equally spaced places on edge of the circle.
2. Place a second flower cage circle on top of it.
3. Place some glue in the middle of the second circle.
4. Place a third circle on top of the second circle.
5. Continue gluing on circles alternating the glue locations.
6. Let the glue dry. Pull apart the circles and watch the garland unfold. Hang with string.

■ Paper Weaving
[Makes 1]

NOTE: This takes some patience. The weaving should be attractive on both sides. The amount of paper depends on the project. A placemat could take quite a few strips of paper. A Christmas tree decoration could take only a few. Animals can be created from paper weaving. Just add wings or fins or feet.

Materials

2 9 inch by 12 inch pieces of construc-   ruler
     tion paper, different colors   scissors
pencil   glue

Procedure

1. Fold 1 sheet of construction paper in half horizontally.
2. Use ruler and pencil to make 1-inch border on 9-inch side. This is the cutting limit.
3. From the fold make cuts to the line. Each cut should be about an inch from the next cut.
4. Unfold the paper. This portion will serve as both loom and warp.
5. Cut the other piece of construction paper into strips 1 inch and 9 inches long. These strips will serve as weft.
6. Weave 1 weft strip over the first warp strip and then under the next strip and then over the next and so on.
7. Weave the next weft strip under the first warp strip and then over the next strip and under the next and so on.
8. Continue weaving until loom is full.
9. Glue edges of warp and weft if desired.

■ Paper Wreath or Star or Flower
[Makes 1]

NOTE: The trick is in the last fold and the glue. A Christmas wreath could be made from green construction paper, and red mistletoe berries (either real or paper) could be added. A star can be made by pinching the ends. Glitter can easily be glued on. Spring flowers could also be made from smaller pieces of construction paper.

Materials

1 piece of construction paper, 12 inches by 18 inches
scissors

glue
stapler and staples

Procedure

1. Fold the paper in half horizontally so that the long ends meet.
2. Fold 1 long edge down about an inch.
3. Fold the other long edge down about an inch but in the other direction.
4. Make quite a few cuts from the middle fold to these new folds. Each new cut should be about 3/4 of an inch away from the previous cut.
5. Unfold paper and reverse paper so that the previous inside is now the outside.
6. Lay one short edge over the other so that a tube is formed. Glue the overlapped edges together.
7. Bring ends of tube together to form wreath and staple 2 ends together.
8. Decorate and enjoy.

■ Origami Fortune Teller
[Makes 1]

NOTE: Also known as a "Cootie Catcher," the fortune teller has been around for almost a hundred years. Children could change the message to create more fun.

Materials

1 piece of square paper at least 8 1/2   markers
     inches by 8 1/2 inches

Procedure

1. Fold the paper in diagonal and make a crease.
2. Open paper and fold opposite diagonal and crease.
3. Unfold paper. Fold each of 4 corners to center of paper (where diagonals cross).
4. Turn paper over and fold those 4 corners to center.
5. Number each of the small triangles.
6. Under each triangle write a different message.
7. Turn fortune teller over and write the name of a color in each corner.
8. Fold fortune teller in half so that the colors are on the outside.
9. Fold fortune teller in half the opposite way. Repeat steps 8 and 9 until fortune teller folds easily.
10. Place thumb and index finger of each hand into the pockets created where the colors are written. You are ready to tell fortunes!

CHAPTER

25
Snow Globes

Snow globes have been around for many years. Children could make snow globes as gifts. Large baby food jars are just about the right size for this project. Children could root through their own toy chests for the toys or figures. Even plastic dinosaurs could be added. Make sure the toys or figures that go inside the snow globes are not made of metal. They would rust in the water. The distilled water helps keep mold away. Consider adding a bit of antiseptic mouthwash to each jar. The antiseptic mouthwash will keep down the growth of microbes.

■ Snow Globe 1 (White Corn Syrup)
[Makes 1]

NOTE: The white corn syrup allows the glitter to drift down slowly.

Materials

1 small, clear jar with tight-fitting lid white corn syrup
1 small toy or figure   glitter or plastic confetti
tacky glue (waterproof glue)

Procedure

1. Glue figure or toy to inside of lid with tacky glue. Let dry overnight.

2. Fill jar almost to the top with corn syrup.

3. Add glitter or plastic confetti.

4. Spread glue around top of jar.

5. Screw lid on and let glue dry.

■ Snow Globe 2 (Glycerin)
[Makes 1]

NOTE: The glycerin changes the water slightly so that the glitter drifts down slower. Glycerin can be purchased at a pharmacy.

Materials

1 small, clear jar with tight-fitting lid distilled water
1 small toy or figure   several drops of glycerin
tacky glue (waterproof)   glitter or plastic confetti

Procedure

1. Glue figure or toy to inside of lid with tacky glue. Let dry overnight.

2. Fill jar almost to the top with distilled water. Add several drops of glycerin.

3. Add glitter or plastic confetti.

4. Spread glue around the top of jar.

5. Screw lid on and let glue dry.

■ Snow Globe 3 (Moth Flakes)
[Makes 1]

NOTE: Moth flakes are added to wool garments to keep moths away. Children could speculate as to the composition of moth flakes and why they do not dissolve in distilled water.

Materials

1 small, clear jar with tight-fitting lid
1 small toy or figure
tacky glue (waterproof)

distilled water
2 teaspoons moth flakes

Procedure

1. Glue figure or toy to inside of lid with tacky glue. Let dry overnight.
2. Fill jar almost to the top with distilled water.
3. Add moth flakes.
4. Spread glue around top of jar.
5. Screw lid on and let glue dry.

■ Snow Globe 4 (Baby Oil)
[Makes 1]

NOTE: The crayon shavings should be brightly colored. The colors add lots of interest to a snow globe.

Materials

1 small, clear jar with tight-fitting lid
1 small toy or figure
tacky glue (waterproof)

baby oil
crayon shavings

Procedure

1. Glue figure or toy to inside of lid with tacky glue. Let dry overnight.
2. Fill jar almost to the top with baby oil.
3. Add crayon shavings.
4. Spread glue around top of jar.
5. Screw lid on and let glue dry.

26
Breads and Baked Goods

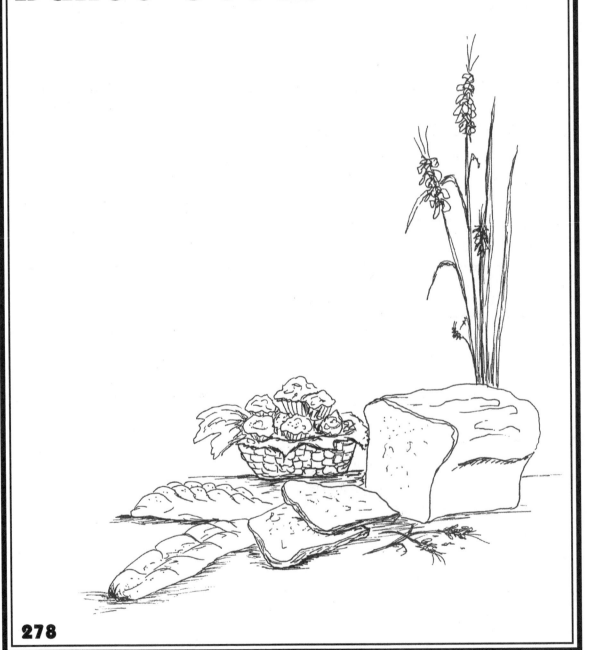

Breads and baked goods rise when leaveners are introduced. The main type of biological leavener is yeast. As yeast grow and multiply, they produce carbon dioxide. The carbon dioxide makes bubbles in the dough, and the dough expands. Yeast breads, obviously, use yeast. However, yeast breads are not the easiest to work with, and they require time and patience.

Chemical leaveners include baking soda and baking powder. They react with liquids to produce carbon dioxide. These leaveners are used in quick breads, muffins, cakes, and cookies.

Children can conduct great science experiments with yeast. Yeast grows best in warm liquids with a bit of sugar. Children could set up several small containers, each containing the same amount of yeast. They can vary the water temperature and the amount of sugar and observe the results. One nice by-product is that the smell is wonderful!

■ Basic Yeast Bread
[Makes 2 loaves]

NOTE: Making yeast bread is a great way to culminate U.S. history studies of the colonial period or the Civil War. Because bread was an important part of the diet during those periods, bread making was an integral household chore. To make whole wheat bread, change the basic recipe by substituting 2 cups of whole wheat flour for white flour. Make rye bread by substituting 3 cups rye flour for the wheat flour and adding 1/4 cup caraway seeds.

Materials

1/4 cup warm water (110°F)
1 package (1 tablespoon) dry yeast
2 tablespoons sugar
2 cups warm milk (110°F)
2 tablespoons butter, melted and cooled
1 teaspoon salt
7 cups all-purpose flour
small mixing bowl
large mixing bowl

wooden spoon
floured surface for kneading
bowl in which to let dough rise
sharp knife
shortening to grease bowl and loaf pans
clean cloth to cover rising bread
2 loaf pans
oven

Procedure

1. Combine water, yeast, and 1 tablespoon sugar in small mixing bowl. Let proof for 15 minutes.

2. Combine milk, melted butter, salt, 1 tablespoon sugar, and yeast mixture in large mixing bowl.

3. Add 3 cups flour, stirring after each cup is added.

4. Beat in 1 cup flour, working the dough until it is smooth and elastic. This takes 5–10 minutes.

5. Add another cup flour, stirring until the dough keeps its shape.

6. Sprinkle 1 cup flour on kneading surface.

7. Remove dough from bowl and place on kneading surface. Knead dough for at least 5 minutes. The longer you knead, even up to 30 minutes, the airier the bread will be.

8. Grease bowl in which bread will rise.

9. Put dough in bowl, and then turn dough over so that the greased surface is on top.

10. Cover bowl with clean cloth. Let rise in a warm area until volume doubles (about 1 hour).

11. Punch dough down, remove from bowl, and knead again for a few minutes.

12. Using sharp knife, divide dough into 2 parts. Form each half into a loaf.

13. Grease loaf pans. Place dough into pans and again cover and let rise until volume doubles (about 45 minutes).

14. Bake at 375°F until loaves are crusty and brown (about 30–45 minutes).

■ Sourdough Starter
[Makes about 3 cups]

NOTE: Sourdough bread was consumed a great deal by early pioneers and miners in the United States because other leavening agents were very hard to obtain. When settlers were ready to begin baking, they took a portion of the starter to add to the bread ingredients. They replenished the starter with flour, water, and sugar. These ingredients kept the starter active. Children could make and then eat sourdough bread after reading *The Bread Sister of Sinking Creek* by Robin Moore.

Sometimes the wrong kind of yeast can make a starter thick and smelly. Dispose of it if this happens. A good sourdough starter should have the consistency of cream and have a wonderful smell!

Materials

1 package (1 tablespoon) dry yeast	ceramic or glass bowl
3 cups flour	cheesecloth
2 cups warm water (110°F)	refrigerator
1 tablespoon sugar	

Procedure

1. Mix flour, yeast, sugar, and warm water in bowl. Cover with cheesecloth.

2. Store in a warm place for 2 days.

3. Place in refrigerator until you are ready to use it.

■ Sourdough Bread
[Makes 3 loaves]

NOTE: Sourdough bread has a sharp, tangy taste. It was a favorite of Yukon prospectors, nicknamed "sourdoughs."

Materials

1 1/2 cups "Sourdough Starter" (see the preceding recipe) at room temperature
3/4 cup milk
2 tablespoons sugar
2 teaspoons salt
2 tablespoons butter, melted
5 cups all-purpose flour
1 package (1 tablespoon) dry yeast
1/2 cup warm water (110°F)

1 tablespoon butter to grease bowl
pan
mixing bowl
mixing spoon
bowl in which to let dough rise
floured surface for kneading
clean cloth to cover dough
cookie sheet
stove with oven

Procedure

1. Scald milk and then stir in sugar, salt, and butter.

2. Combine yeast and warm water in mixing bowl.

3. Add sourdough starter, milk mixture, and 3 1/2 cups flour to mixing bowl.

4. Sprinkle some flour on kneading surface and knead dough. Work in more flour until dough is smooth and shiny.

5. Grease bowl with butter. Put dough into bowl and rub a little butter on top surface of dough.

6. Cover with clean cloth and place in a warm spot. Let dough double in size (about 1 hour).

7. Punch dough down and remove it from bowl. Place dough on kneading surface again. Knead dough again. Divide dough into 3 parts.

8. Shape into loaves and cut 1/2 slit on tops of loaves. Put loaves on cookie sheet.

9. Cover with damp cloth. Let rise until double again.

10. Bake at 400°F for 10 minutes. Then reduce oven temperature to 375°F and bake for 45 minutes to 1 hour.

■ Challah

[Makes 2 loaves]

Challah is often served in Jewish household on Friday nights. It can also be part of "breakfasts" for Jewish holidays such as Yom Kippur. This slightly sweet bread can be fashioned into a braid or into a round loaf.

Materials

1 package (1 tablespoon) dry yeast	large mixing bowl
1/2 cup warm water (110°F)	mixing spoon
7 cups all-purpose flour	bowl in which to let dough rise
1/2 teaspoon salt	floured surface for kneading
1 cup hot water	clean cloth to cover rising bread
1/3 cup vegetable oil	parchment paper
1 tablespoon honey	cookie sheet
4 large eggs	small mixing bowl
1 tablespoon water	pastry brush
vegetable oil to coat bowl	oven

Procedure

1. Combine yeast with 1/2 cup warm water in large mixing bowl. Let proof for 15 minutes.

2. Add salt and 5 1/2 cups flour.

3. Add 1 cup hot water, vegetable oil, honey, and 3 large eggs.

4. Turn out onto kneading surface and knead for 5 minutes. Add extra flour if dough sticks to surface. Dough should be stiff and pliable at the end of kneading.

5. Grease bowl with vegetable oil. Place dough in bowl. Turn dough in bowl so that oil coats top surface of dough.

6. Cover with clean cloth and place in a warm spot. Let double in size (about 30 minutes).

7. Return dough to kneading surface and punch down. Divide dough into 2 parts. Divide each part into 3 sections.

8. Roll each section into a log 15 inches long. Braid 3 logs together. Close ends. Repeat steps to make other loaf.

9. Line cookie sheet with parchment paper. Place loaves on paper. Cover with clean cloth and let double in volume (about 30 minutes).

10. Stir remaining egg and 1 tablespoon water in small mixing bowl. Use pastry brush to brush loaf surfaces with this egg mixture.

11. Bake at 350°F for 45 minutes. Tops should be crusty and brown.

■ Soft Pretzels

[Makes 15 pretzels]

NOTE: Soft pretzels, the pride of Philadelphia, are often served with mustard. Although children can make traditional pretzel shapes, the coils of dough can become expressions of creativity. This recipe takes 2 days.

Materials

1 package (1 tablespoon) yeast	mixing bowl
1 cup warm water (110°F)	mixing spoon
1/4 cup sugar	2 small bowls
1/2 teaspoon salt	small, covered container
2 tablespoons shortening	clean cloth to cover rising dough
1 large egg, room temperature	nonstick cooking spray
3 1/2 cups all-purpose flour	cookie sheet
coarse salt	refrigerator
mustard (optional)	oven

Procedure

1. Pour warm water into mixing bowl. Sprinkle yeast over water. Stir to dissolve.

2. Add sugar, salt, and shortening.

3. Separate the egg, leaving the yolk in one small bowl and the white in the other small bowl.

4. Combine half the egg yolk with 1 tablespoon water. Cover container and store until almost the last step.

5. Add egg white and other half of egg yolk to yeast mixture.

6. Slowly add 2 cups flour, stirring constantly. Add more flour until dough is stiff. Cover and refrigerate until next day.

7. Divide dough into 15 pieces. Make coils of dough and fashion into standard pretzel shapes or original designs.

8. Spray cookie sheet with nonstick cooking spray. Place pretzels on cookie sheet.

9. Brush tops with egg yolk–water mixture. Sprinkle with coarse salt.

10. Cover and let rise until double in volume (about 45 minutes).

11. Bake at 400°F for 12 minutes or until nicely brown. Serve with mustard, if desired.

■ Irish Soda Bread
[Makes 2 loaves]

NOTE: Make Irish soda bread to celebrate Saint Patrick's Day. The loaves could be topped with a bit of green icing. Irish soda bread uses both baking powder and baking soda as leaveners.

Materials

4 1/2 cups all-purpose flour
1 teaspoon salt
3 teaspoons baking powder
1 teaspoon baking soda
1/4 cup sugar
1/4 cup butter
pastry cutter
1 large egg
1 3/4 cups buttermilk

2 mixing bowls
mixing spoon
floured surface for kneading
nonstick cooking spray
2 pie plates or cake pans (8-inch diameter)
knife
oven

Procedure

1. Combine flour, salt, baking powder, baking soda, and sugar in a bowl. Cut in butter with pastry cutter. Mixture should be crumbly.

2. In other bowl, stir together egg and buttermilk.

3. Stir liquid ingredients into dry ingredients.

4. Turn dough onto kneading surface and knead for 2–3 minutes. Make 2 round loaves from dough.

5. Spray each pan with nonstick cooking spray. Place dough into pans. Shape dough to fit pans.

6. Cut a cross (1/2 inch deep) on each loaf. Bake at 375°F for 35–40 minutes or until brown.

■ Basic Quick Bread or Muffin Mix
[Makes 1 loaf of bread or 12 muffins]

NOTE: This versatile mix allows for many variations. Make muffins after reading *If You Give a Moose a Muffin* by Laura Joffe Numeroff.

Materials

2 large eggs
3/4 cup milk
1/2 cup vegetable oil
2 cups all-purpose flour
1/3 cup sugar
3 teaspoons baking powder

1 teaspoon salt
mixing bowl
mixing spoon
1 loaf pan or 1 muffin pan (12 cup)
nonstick cooking spray
oven

Procedure

1. Beat eggs in mixing bowl and add milk and vegetable oil.
2. Stir in remaining ingredients.
3. Spray pan with nonstick cooking spray. Add dough.
4. Bake 1 loaf bread at 375°F for 50–60 minutes, or 12 muffins at 400°F for 20 minutes.

■ Fruit Bread
[Makes 1 loaf]

NOTE: To the "Basic Quick Bread" recipe (above) add 1/2 to 1 cup fresh, frozen, or dried fruits, such as blueberries, apples, raisins, or chopped apricots.

■ Orange–Honey Bread
[Makes 1 loaf]

NOTE: Replace sugar with honey in the "Basic Quick Bread" recipe (above). Add 2 tablespoons grated orange zest to batter. Spread 1/2 cup orange marmalade on top of bread or muffins when fresh from oven.

■ Hearty Grain–Raisin Bread
[Makes 1 loaf]

NOTE: Substitute 1 cup quick-cooking oats, granola, wheat germ, or whole wheat for 1 cup flour in the "Basic Quick Bread" recipe (above). Add 1 cup raisins, 1/2 teaspoon ground nutmeg, and 1/2 teaspoon cinnamon. Use brown sugar instead of regular sugar.

■ Biscuits
[Makes 9 biscuits]

NOTE: Serve these biscuits warm with jam. The addition of raisins and cinnamon gives the biscuits pizzazz. Children could make these biscuits after reading any of the *Biscuit* books written by Alyssa Satin Capucilli and illustrated by Pat Schories.

Materials

2 cups all-purpose flour
1 1/2 teaspoons baking powder
1/2 teaspoon salt
2 tablespoons shortening
pastry cutter
3/4 cup milk
mixing bowl

mixing spoon
square baking pan 8 inches by 8
   inches
nonstick cooking spray
oven
knife

Procedure

1. Combine dry ingredients in mixing bowl.
2. Cut in shortening using pastry cutter.
3. Add milk.
4. Spray baking pan with nonstick cooking spray. Pour dough into baking pan. Evenly distribute dough.
5. Bake at 375°F for 20 minutes. Cool and slice into squares with knife.

■ Shortcake
[Makes 9 shortcakes]

NOTE: To the "Biscuits" recipe (above) add 2 tablespoons sugar to dry ingredients. After placing dough in pan, sprinkle surface with 1 tablespoon sugar. Bake as per "Biscuits" recipe.

After baking, allow shortcake to cool. Cut into squares and place on plates. Top with peaches, strawberries, blueberries, or other fruit. Add some whipped cream.

■ Scones
[Makes 16 scones]

NOTE: Scones are closely related to biscuits. Most scones are a bit sweeter than biscuits. Also, some scone recipes include eggs, but biscuit recipes are egg-free. Make and serve scones after reading something very British, like *The Lion, the Witch and the Wardrobe* by C.S. Lewis.

Materials

2 cups all-purpose flour
2 teaspoons baking powder
1/2 teaspoon baking soda
1/2 cup (1 stick) butter, cut into small pieces
2 tablespoons sugar
1 large egg
3/4 cup buttermilk
2 tablespoons sugar to sprinkle on top of dough surface

mixing bowl
mixing spoon
pastry cutter
floured surface for kneading
knife
baking sheet
oven

Procedure

1. Combine dry ingredients in mixing bowl.
2. Use pastry cutter to cut butter into dry ingredients.
3. Add buttermilk and stir until dough forms a ball.
4. Turn dough onto floured surface and knead for 1 minute.
5. Divide dough into 2 parts. Form each half into a circle with a diameter of 6 inches.
6. With knife, cut each circle into 8 wedges. Sprinkle each wedge with a bit of sugar.
7. Carefully transfer each wedge onto baking sheet.
8. Bake at 375°F for 18–20 minutes.

■ Hardtack
[Makes 15 large hardtacks]

Sixteenth-century sailors lived on a diet of primarily salt beef and hardtack (dry biscuits.) Hardtack was also an important food for soldiers during the Civil War. Diaries indicate hardtack was eaten even when infested with insects. This recipe comes from a Civil War buff.

Materials

4 cups all-purpose flour
2 teaspoons salt
1/2 cup shortening
1 cup water
mixing bowl
mixing spoon
pastry cutter
floured surface for kneading

rolling pin
knife
spatula
baking sheets
very clean skewer or clean handle
   from watercolor brush
oven

Procedure

1. Combine flour and salt together in mixing bowl.
2. Using pastry cutter, cut in shortening.
3. Stir in water.
4. Turn dough onto floured surface and knead until dough resembles clay. Roll into a 1/2-inch thick sheet.
5. Cut into 3 inch by 3 inch squares.
6. With spatula, transfer hardtack to baking sheets.
7. With skewer, make 4 rows of 4 holes each in each square. Make sure holes go completely through hardtack.
8. Bake at 400°F for 40 minutes or until golden brown.
9. Cool and serve. Store any remaining hardtack in an airtight container.

■ Corn Bread
[Makes 9 servings]

NOTE: Native Americans introduced a kind of corn bread to early settlers.

Materials

1 1/2 cups yellow cornmeal
1 cup all-purpose flour
1/3 cup sugar
1 tablespoon baking powder
1/2 teaspoon salt
1 1/2 cups milk
3/4 cup butter, melted and cooled
2 large eggs

mixing bowl
spoon
nonstick cooking spray
baking pan, 9 inches by 5 inches by
   3 inches
oven
knife

Procedure

1. Combine all ingredients in mixing bowl.

2. Spray baking pan with nonstick cooking spray. Pour in batter.

3. Bake at 400°F for 35 minutes.

4. Cool and cut into squares.

■ Corn Tortillas
[Makes 15 tortillas]

NOTE: Tortillas can be eaten alone. However, they are usually the base for such dishes as tacos, burritos, and enchiladas. A tortilla is a flat bread. It has no yeast or other leavening.

Materials

2 1/2 cups corn flour (also known as masa harina, available in grocery stores and gourmet shops)
1/2 teaspoon salt
1 1/2 cups warm water (110°F)
mixing bowl

mixing spoon
floured surface for kneading
wax paper
rolling pin
electric frying pan
spatula

Procedure

1. Combine corn flour and salt in mixing bowl. Add 1 cup warm water and stir. If necessary, add remaining water to make a dough.

2. Turn dough onto floured surface and knead. Add more corn flour if dough is too sticky.

3. Divide dough into 15 pieces.

4. Place a piece between sheets of wax paper. With rolling pin, roll dough into a flat, circular shape 4 inches in diameter. Repeat process until all 15 tortillas are made.

5. Place a tortilla in frying pan and cook at a high temperature for 2 minutes. Flip tortilla with spatula and brown the other side. Repeat process until all 15 tortillas are cooked.

■ Journey Cakes (also called Johnny Cakes)
[Makes 15 small cakes]

NOTE: Both Native Americans and pioneers ate journey cakes. Nutritious and easy to make, the bread traveled well. Original recipes used dried corn. This adaptation uses cornmeal. The final product resembles a pancake and can be served with jam or honey.

Materials

1 1/4 cups milk

2 large eggs

1 cup cornmeal

1 teaspoon sugar

mixing bowl

mixing spoon

electric frying pan

nonstick cooking spray

spatula

jam or honey (optional)

Procedure

1. Combine milk and large eggs in mixing bowl.

2. Add dry ingredients.

3. Coat cooking surface of electric frying pan with nonstick cooking spray and heat frying pan to 350°F.

4. Drop 4 small amounts of batter into frying pan.

5. Let first side brown. Using spatula, flip cakes onto other sides and brown.

6. Remove journey cakes from frying pan.

7. Repeat steps 4 through 6 until all journey cakes are prepared.

8. Serve with jam or honey, if desired.

■ Basic Cookies
[Makes 6 dozen]

NOTE: Few foods can beat the smell and taste of freshly baked cookies. This recipe can be adapted in many ways. The dough can even be rolled out and shaped with cookie cutters. One group of children I know bakes heart-shaped cookies around Valentine's Day. At lunch they sell the cookies to friends, and the proceeds go to the American Heart Association.

Materials

3/4 teaspoon salt
1 teaspoon baking powder
1/4 teaspoon baking soda
3 3/4 cups all-purpose flour
3/4 cup (1 1/2 sticks) butter
2/3 cup shortening
1/2 cup sugar
1 cup brown sugar
2 large eggs
2 1/2 teaspoons vanilla

nonstick cooking spray
medium-size mixing bowl
large mixing bowl
mixing spoon
wax paper
knife
cookie sheets
refrigerator
oven

Procedure

1. Combine salt, baking powder, baking soda, and flour in medium-size mixing bowl.

2. In large mixing bowl, blend butter and shortening. Slowly add sugar, brown sugar, eggs, and vanilla.

3. Stirring constantly, add dry ingredients to shortening–sugar mixture.

4. Divide dough into thirds. Shape each portion into a log. Wrap each log in wax paper and refrigerate overnight. Or, the dough can be frozen for later use.

5. Remove dough from refrigerator and take off wax paper. Cut dough into 1/2-inch thick slices.

6. Spray nonstick cooking spray onto cookie sheets.

7. Place cookies onto cookie sheets.

8. Bake at 400°F for 6 minutes.

■ Chocolate Chip Cookies
[Makes 6 dozen]

NOTE: Using the "Basic Cookies" recipe (p. 292), add 1 1/2 cups semisweet chocolate chips to the shortening–sugar mixture before adding dry ingredients.

■ Chocolate Cookies
[Makes 6 dozen]

NOTE: Using the "Basic Cookies" recipe (above) add 4 squares unsweetened chocolate (melted) to the shortening–sugar mixture before adding dry ingredients.

■ Sugar Cookies
[Makes 6 dozen]

NOTE: Using the "Basic Cookies" recipe (above) sprinkle colored sugar on cookies before baking.

■ Snickerdoodles
[Makes 6 dozen]

NOTE: Snickerdoodles are drop cookies. They require less work and time than sliced cookies.

Procedure

1. Follow "Basic Cookies" recipe (above) through step 3. Do not slice the dough but drop the dough by spoonfuls onto cookie sheets.
2. Combine 1/3 cup sugar with 2 tablespoons cinnamon. Sprinkle on cookies before baking.

■ Butterscotch Cookies
[Makes 6 dozen]

NOTE: Use the "Basic Cookies" recipe (above). However, replace the cup of sugar with another cup of brown sugar.

■ Fortune Cookies
[Makes 30 cookies]

NOTE: Fortune cookies probably did not originate in China. However, these cookies make a great treat on Chinese New Year. Bake 4 cookies at a time. When they come out of the oven, they are flexible enough to fold. After sitting for a few minutes, the cookies harden and will break rather than fold. If they do harden before they can be bent, return them to the oven for 20 seconds.

Materials

1 cup all-purpose flour
2 tablespoons cornstarch
1/2 cup sugar
1/2 cup vegetable oil
4 egg whites (1/2 cup)
2 teaspoons grated orange zest
1 tablespoon orange flavoring
nonstick cooking spray

mixing bowl
mixing spoon
electric mixer
cookie sheets
spatula
30 fortunes, typed on slips of paper
paper cups
oven

Procedure

1. In mixing bowl, combine flour, cornstarch, and sugar.
2. Add vegetable oil and egg whites. Beat at high speed until batter is well blended.
3. Add grated orange zest and orange flavoring.
4. Spray cookie sheets with nonstick cooking spray. Drop level tablespoons of batter on cookie sheets.
5. Use spatula to spread batter into a 4-inch diameter circle. Bake 4 cookies at a time.
6. Bake at 325°F for 10 minutes, or until cookies are lightly browned.
7. Remove cookies from cookie sheet with spatula.
8. Place a fortune in the center of each cookie. Fold cookies in half.
9. Fold cookies in half again by creasing the folded edge on the edge of the mixing bowl.
10. To make sure the cookie retains its shape, place folded cookie in a paper cup with the points down. Let cookies harden for 2 minutes. Remove them from cups.
11. Repeat process with the rest of the cookie batter and bake the cookies.
12. Store cookies in an airtight container.

■ Chocolate Crisp Cookies
[Makes 3 dozen]

NOTE: Because chocolate crisp cookies require no baking, they are quick and easy to make.

Materials

8-ounce milk chocolate bar
1/3 cup shredded coconut
2 cups crisped rice cereal
microwave-safe bowl

mixing spoon
aluminum foil
microwave oven
refrigerator

Procedure

1. Microwave chocolate bar in microwave-safe bowl at medium setting for 1 minute. Chocolate bar should be completely melted.

2. Add coconut and crisped rice cereal.

3. Drop by spoonfuls on aluminum foil.

4. Refrigerate for 1 hour.

■ Rugela

[Makes 32–36 cookies]

NOTE: Also spelled rugala and rugelach, these cookies are made all year long. However, they are very popular at Hanukkah.

Materials

Dough:

2 1/3 cups all-purpose flour

1/2 pound (2 sticks) butter, softened

1 cup sour cream

mixing bowl

mixing spoon

plastic wrap

extra flour

rolling pin

working surface with extra flour

knife

cookie sheets

nonstick cooking spray

Filling:

1/3 cup sugar

1 teaspoon sugar

1 teaspoon cinnamon

2 tablespoons finely chopped nuts.

small mixing bowl

mixing spoon

Procedure

1. Combine 2 cups flour, butter, and sour cream in mixing bowl.

2. While stirring, add the last 1/3 cup of flour.

3. Shape into a ball and cover with plastic wrap.

4. Refrigerate overnight.

5. Remove from refrigerator and divide into 4 portions.

6. Lightly dust working surface and place one of the 4 portions on surface.

7. Roll out into a circle about 12 inches in diameter.

8. Combine ingredients for the filling in small mixing bowl.

9. Spread 1 tablespoon of filling on the dough circle.

10. Cut dough like a pizza into 16 triangles.

11. Start at the wide point of each triangle and roll to point.

12. Repeat the process with the other pieces of dough.

13. Spray cookie sheets with nonstick cooking spray.

14. Place rugela on cookie sheets and bake at 350°F for about 15 minutes or until golden brown.

■ Hamantaschen
[Makes about 60 cookies]

NOTE: Hamantaschen are eaten during Purim, a Jewish celebration. Esther, the Jewish wife of a king saved the Jews from being destroyed by the evil advisor Haman. Purim occurs on the 14th or 15th day of the Jewish month Adar. Adar roughly corresponds to late February or March.

Materials

5 large eggs
1 1/2 cups sugar
1 cup vegetable oil
1/2 cup orange juice
grated zest of 1 orange
grated zest of 1 lemon
1 tablespoon lemon juice
1 teaspoon vanilla
6 1/2 cups all-purpose flour
1 1/2 teaspoons baking powder
1/2 teaspoon salt

3 cups fruit filling
2 large mixing bowls
whisk
mixing spoon
parchment paper
baking trays
rolling pin
working surface with extra flour
cookie cutter with 3-inch diameter
refrigerator

Procedure

1. In a large mixing bowl beat eggs with whisk.
2. Slowly add sugar, stirring constantly.
3. Add vegetable oil, orange juice, lemon juice, vanilla, and grated zests of orange and lemon.
4. In other mixing bowl combine flour, baking powder, and salt.
5. Slowly stir dry ingredients into egg mixture.
6. Spread dough on parchment-lined baking trays.
7. Refrigerate for at least 3 hours.
8. Remove 1/4 of the dough from the refrigerator.
9. Dust working surface with flour and place dough on top.
10. Knead until smooth.
11. Roll out the dough with the rolling pin until dough is 1/4 inch thick.
12. Use cookie cutter to make circles of dough.
13. Place circles close together on a baking tray.
14. Reroll scraps and make more circles.
15. Place a teaspoon of fruit filling into the center of each circle.
16. Fold over three sides of the dough around the filling.
17. Pinch dough to make triangles. Some of the fruit filling should be seen.
18. Repeat the process with the rest of the refrigerated dough.
19. Bake at 350°F for 20–25 minutes or until the cookies are nicely brown.

■ Easy Doughnuts
[Makes 24]

NOTE: Children could make a "doughnut" chart (like a pie chart) of their favorite kinds of doughnuts.

Materials

3 cans 8-count refrigerated biscuits
knife
vegetable oil
frying pan with heating element or
  electric skillet

resealable plastic bags
confectioners' sugar

Procedure

1. Remove biscuits from cans and separate.
2. Use the knife to make a small hole in each biscuit.
3. Heat oil in the frying pan to about 375°F.
4. Carefully drop biscuits into the oil and fry.
5. Remove one at a time and let cool.
6. Pour confectioners' sugar into plastic bags.
7. Drop each doughnut one at a time into the confectioners' sugar and toss until coated.
8. Remove and enjoy.

■ Doughnuts from Scratch
[Makes about 30 doughnuts]

NOTE: Doughnuts may not be nutritious, but they are good comfort food. Children could read Robert McCloskey's *Homer Price*. The book contains 6 stories, and one of those stories concerns a doughnut-making machine.

Materials

2 large eggs
1 cup sugar
2 tablespoons (1/4 stick) melted butter
5 cups all-purpose flour
3 teaspoons baking powder
3/4 teaspoon nutmeg
1/4 teaspoon salt
1 cup milk
oil for deep-fat fryer
confectioners' sugar (as optional topping)
mixture of 1 1/2 teaspoons cinnamon and 2 tablespoons sugar (as optional topping)
large bowl
smaller bowl
mixing spoon
deep-fat fryer
large slotted spoon
paper towels

Procedure

1. Beat eggs in large bowl and add melted butter and sugar.
2. Combine flour, baking powder, salt, and nutmeg in the smaller bowl and then add to larger bowl.
3. Add milk and combine.
4. Form 30 doughnuts.
5. Pour oil into deep-fat fryer. Heat deep-fat fryer to 375°F. Lower each doughnut gently into oil with large slotted spoon.
6. The doughnuts will rise to the surface of the oil when they are almost done (about a minute). Cook a bit more. They should be golden brown.
7. Remove with the large slotted spoon and place on paper towels to drain.
8. Doughnuts may be sprinkled with confectioners' sugar or cinnamon–sugar mixture.

■ Welsh Griddle Cakes
[Makes about 24]

NOTE: These are best eaten right away. That is usually not a problem!

Materials

1 cup sugar	mixing bowl
4 cups all-purpose flour	mixing spoon
3 teaspoons baking powder	rolling pin
1/2 teaspoon salt	floured working surface and extra flour
1 teaspoon nutmeg	
3 sticks butter	3-inch cookie cutter
1 cup currants or raisins	griddle
2 large eggs	heating element
2 tablespoons milk (if necessary)	cooling rack
1 cup superfine sugar (for dusting)	

Procedure

1. Combine sugar, flour, baking powder, salt, nutmeg, and butter in mixing bowl.
2. Stir in currants or raisins.
3. Mix in 2 large eggs.
4. A stiff dough should result. If necessary, add a bit of milk.
5. Roll dough out on the floured surface until the dough is about 1/4 inch thick.
6. Use cookie cutter to make rounds. Gather scraps and roll again until almost all the dough has been used.
7. Heat griddle to a medium heat. Bake cakes on 1 side. They should rise a bit and become shiny.
8. Turn them over and bake the other sides.
9. Place on cooling racks and dust with superfine sugar.

■ Crumpets
[Makes 10–12]

NOTE: This traditional English bread is served with butter. However, it is fried and not baked.

Materials

1 packet (1 tablespoon) yeast
1 large egg
2 cups warm milk (110°F)
1/4 teaspoon salt
1 ounce butter
4 cups all-purpose flour
2 small mixing bowls
1 large mixing bowl
whisk
2 mixing spoons

wax paper
kneading surface with extra flour
crumpet ring or clean tuna fish can with both end metal pieces removed
frying pan
heating element
extra butter for frying
spatula

Procedure

1. Mix yeast with 1/2 cup warm milk in small mixing bowl.
2. In large mixing bowl combine the rest of milk with butter.
3. With the whisk beat egg in small mixing bowl and then add to milk–butter mixture.
4. Add salt and 3 cups flour. Mix well.
5. Add yeast mixture and mix well again. Add more flour if necessary to form a dough.
6. Cover dough with wax paper and allow dough to rest in a warm location for 20 minutes.
7. Make 10–12 balls of dough and mold in the crumpet ring or tuna fish can "ring."
8. Place crumpets on wax paper and allow to rise for 15–20 minutes.
9. Heat frying pan and add butter. Fry each crumpet for a couple of minutes on each side.
10. Serve warm. Split each crumpet and spread with butter.

■ Grandma's Buttermilk Pancakes
[Makes 12–14 pancakes]

NOTE: This recipe has been in our family for generations. Neighborhood kids would come help us fix these pancakes. To make buttermilk waffles, decrease buttermilk and batter will be thicker.

Materials

2 cups all-purpose flour
1 teaspoon soda
1 teaspoon salt
1 tablespoon sugar
2 large eggs
about 1 cup buttermilk

mixing bowl
mixing spoon
frying pan
nonstick cooking spray
spatula

Procedure

1. Combine flour, soda, salt, and sugar in mixing bowl.

2. Add eggs and mix well.

3. Add about 1/2 cup buttermilk and combine. Add more buttermilk as necessary to make a thick batter.

4. Heat frying pan and coat bottom of frying pan with nonstick cooking spray.

5. Pour the pancake batter. Flip them over when edges are brown and surface bubbles have broken.

6. Serve with butter and warm syrup.

■ Pita (Flat Bread)
[Makes 12]

NOTE: Pita may be the earliest type of bread to be made. Experts believe it started in the eastern Mediterranean region.

Materials

1 envelope (1 tablespoon) yeast
1 1/4 cups lukewarm water
2 teaspoons honey
1 teaspoon salt
3 cups all-purpose flour
1/2 stick butter

large bowl
mixing spoon
kneading surface and extra flour
cookie sheets
damp cloth

Procedure

1. In large bowl combine yeast, water, and honey. Let mixture sit in a warm place for about 5 minutes or until it begins to bubble.
2. Stir in salt and flour.
3. Knead mixture on kneading surface for 10 minutes.
4. Divide dough into 12 pieces.
5. Pat each piece into a circle about 6 inches across and 1/4 inch thick.
6. Butter cookie sheets.
7. Lay pitas on buttered cookie sheets and cover with a damp cloth. Let rise in a warm place for about 45 minutes. They will puff up.
8. Remove the damp cloth and place in an oven.
9. Bake at 500°F for about 12 minutes. They should be lightly browned.

■ Chappatis (Fried Bread from Southeast Asian Cultures)
[Makes 50 small chappatis]

NOTE: Chappatis are sometimes served as an alternative to rice. They accompany many dishes and can be frozen.

Materials

2 cups whole wheat flour
2 cups unbleached white flour
1 teaspoon salt
3 tablespoons oil or melted butter
2 cups water
butter for frying

mixing bowl
mixing spoon
kneading surface with extra flour
damp cloth
frying pan

Procedure

1. Combine flours, salt, and oil.
2. Add enough water to make a stiff dough.
3. Knead for 5–10 minutes and then place the dough back in the bowl.
4. Cover dough with a damp cloth and set it aside for an hour.
5. Shape dough into balls the size of large marbles and then flatten each portion.
6. Fry each chappati over medium heat in a small amount of butter.

■ Crepes
[Makes 12 crepes]

NOTE: A little crepe batter goes a long way.

Materials

1 cup all-purpose flour

1/4 teaspoon salt

1 cup milk

3 large eggs

mixing bowl

whisk

mixing spoon

crepe pan or frying pan

stove or heating element

oil

spatula

plate

sugar or jams (optional)

Procedure

1. With the whisk beat together eggs, milk, and salt.

2. Add flour, several teaspoons at a time. Beat until smooth and then add more flour. After all the flour has been added, the batter should be the consistency of light cream.

3. Let batter rest for an hour so that flour can hydrate.

4. Heat crepe pan or frying pan on high and add a small amount of oil. Turn down the heat to about medium.

5. Pour in about 1/4 cup batter. Tilt pan so that batter spreads evenly across the pan.

6. When the edges brown, turn crepe with the spatula.

7. Cook the other side and remove to a plate.

8. Cook the remaining crepes in the same way.

9. Sprinkle sugar or spread jam on each crepe and roll up.

10. Serve warm.

■ Navajo Fry Bread
[Makes 8 servings]

NOTE: Shortening could be substituted for the lard.

Materials

3 1/4 cups all-purpose flour
1 cup nonfat dry milk
1 tablespoon baking powder
1/2 teaspoon salt
10 ounces lard
1 cup ice water
1 teaspoon salt
mixing bowl
pastry blender

mixing spoon
towel
rolling pin
working surface with extra flour
knife
frying pan
heating element
spatula

Procedure

1. Combine flour, nonfat dry milk, baking powder, and salt in mixing bowl.
2. With pastry blender cut in about 5 tablespoons lard. The mixture will resemble cornmeal.
3. Pour in ice water and mix until the dough forms a large ball.
4. Cover with towel and let stand at room temperature for about 2 1/2 hours.
5. Divide dough into 2 portions. Shape each piece into a ball and roll out on the floured surface until each is about 1/4 inch thick.
6. Make about 3 slits partway through the dough and let it rest again for about 10 minutes.
7. Heat the frying pan over medium-high heat and add some of the lard.
8. Place 1 of the dough circles into the hot melted lard and fry for about 4–5 minutes.
9. Flip the dough on the other side and fry again for about 4–5 minutes.
10. Remove bread and drain on paper towels. Fry the other round.
11. Sprinkle both breads with salt and keep them warm. Cut into wedges when they are being served.

■ Sopapilla
[Makes 24]

NOTE: Sopapilla are served as the end to a Mexican dinner in the American Southwest.

Materials

2 cups all-purpose flour
1 teaspoon sugar
1 tablespoon baking powder
1 large egg, beaten
1 teaspoon butter
1 cup milk
frying oil
mixing bowl
mixing spoon

working surface with extra flour
wax paper
rolling pin
frying pan
spatula
paper towels
honey or cinnamon–sugar mixture
    (optional)

Procedure

1. Combine flour, sugar, baking powder, and salt in mixing bowl.
2. Add egg, butter, and milk. Stir until mixture looks like bread dough.
3. Knead well on floured surface for about 5 minutes.
4. Return dough to mixing bowl and cover with wax paper. Let stand for 2 hours.
5. Remove dough from bowl and roll out to 1/4 inch thickness.
6. Cut dough into 4 inch squares.
7. Fry dough in hot oil for about a half minute per side. They cook quickly, and they puff up.
8. Drain on paper towels.
9. Serve hot with honey or cinnamon–sugar mixture.

■ Cinnamon Crisps
[Makes 48]

NOTE: This treat is fairly healthy.

Materials

1/4 cup sugar
2 teaspoons cinnamon
6 flour tortillas
3 tablespoons butter
mixing bowl
mixing spoon

microwave-safe bowl
microwave oven
basting brush
knife
cookie sheet
oven

Procedure

1. Combine sugar and cinnamon in the mixing bowl.

2. Melt butter in the microwave-safe bowl in the microwave oven.

3. Brush both sides of each tortilla with butter. Sprinkle on cinnamon–sugar mixture.

4. Cut each tortilla into 8 pieces and place on cookie sheet.

5. Bake at 400°F for 5–7 minutes or until crisp.

27
Edible Art

■ Chocolate Clay
[Makes about 3/4 cup—enough for 3 children]

NOTE: What can we say? Children can have their clay and eat it, too!

Materials

10 ounces chocolate chips or discs	mixing spoon
1/3 cup corn syrup	wax paper
microwave-safe bowl	tray
microwave oven	

Procedure

1. Pour chocolate chips into microwave-safe bowl and microwave until chocolate melts (about 2–3 minutes).

2. Add corn syrup and mix.

3. Place wax paper on tray.

4. Pour chocolate mixture on wax paper and spread. Cover loosely with more wax paper and allow to stiffen (about an hour)

5. Children can sculpt animals or flowers or geographic features or anything else they can image.

6. Then they can eat their creations!

■ Peanut Butter Clay

[Makes 4 cups—enough for 4 children]

NOTE: Children can make great animals with this concoction. They can add hair (shredded coconut), eyes (raisins), a mouth (sliver of red apple), and many other details. Children with nut allergies should not play with this clay. This process takes 2 days.

Materials

2 cups smooth peanut butter
1 cup honey
3 cups instant dry milk
mixing bowl
mixing spoon

wax paper
extra edibles, such as raisins, shredded coconut, chocolate chips (optional)
refrigerator

Procedure

1. Combine peanut butter and honey in mixing bowl.

2. Add dry milk a bit at a time to make a stiff dough.

3. Refrigerate overnight.

4. The next day remove from refrigerator and divide into portions. Mold on wax paper, decorate, and then eat!

■ Granola Clay
[Makes 3 cups—enough for 3 children]

NOTE: The granola imparts a hearty, substantial feeling to this edible clay.

Materials

1/2 cup maple-flavored syrup
1 cup creamy peanut butter
2 cups instant dry milk
1 cup granola

mixing bowl
mixing spoon
wax paper

Procedure

1. Combine maple-flavored syrup and peanut butter in mixing bowl.

2. Add instant dry milk and granola.

3. Mold on wax paper. Children can make great snowmen from this edible clay. Then they can eat their creations!

■ Marzipan
[Makes 1 1/2 cups—enough for 2 children]

NOTE: This confection dates back to at least medieval times, when a feast often concluded with large marzipan sculptures.

Materials

1 can (8 ounces) almond paste (found in baking section of grocery store)
1 cup confectioners' sugar
2 tablespoons light corn syrup

mixing bowl
mixing spoon
wax paper
food coloring and paint brushes

Procedure

1. Break almond paste into smaller pieces and place in mixing bowl.

2. Gradually stir in confectioners' sugar and light corn syrup.

3. Turn onto wax paper and knead until smooth.

4. Break into pieces so that each child can mold and sculpt.

5. Let shapes dry.

6. Children can then paint their candy sculptures with food coloring.

7. Children can eat their creations!

■ Frosting Clay

[Makes 2 cups—enough for 6 children]

NOTE: Children love to "dig into" this clay. It has a wonderful consistency.

Materials

1 container prepared frosting   mixing spoon
1 cup confectioners' sugar   wax paper
mixing bowl

Procedure

1. Empty frosting into mixing bowl.
2. Slowly add confectioners' sugar and mix until a stiff dough appears.
3. Give each child a dab and let them mold on wax paper. Then they can eat their creations!

■ Cheese Clay

[Makes 2 1/2 cups—enough for 5 children]

NOTE: Because of the orange color of the cheese, this clay makes great Halloween monsters and such.

Materials

10 crackers   rolling pin
1 small container of processed cheese   mixing bowl
    spread   mixing spoon
1/2 gallon plastic bag   wax paper

Procedure

1. Place crackers in plastic bag and close.
2. Using rolling pin, crush crackers into crumbs.
3. Empty container of cheese into mixing bowl.
4. Add enough cracker crumbs to form a good clay.
5. Give each child a dab and let them mold on wax paper. Then they can eat their creations!

■ Edible Finger Paints

NOTE: Young children can practice writing their letters in a very fun, edible way!

Materials

large graham cracker squares
spoons
- prepared puddings
- cheese spreads
- peanut butter

wax paper
any of the following "paints"
- whipped cream
- fruit-flavored yogurt
- chocolate syrup

Procedure

1. Cover work area with wax paper.
2. Give a large graham cracker square to each child.
3. Dab a spoonful of one of the "paints" onto the graham cracker.
4. Let children finger paint on their crackers.
5. Then let them eat their creations.

■ Mashed Potato Geography Creations

NOTE: This goo enlivens a lesson on geography and land features. Children could make mountains, valleys, deltas, and so much more.

Materials

real or instant warm mashed potatoes
warm gravy

forks and spoons
plates

Procedure

1. Give each child a plate, fork, spoon, and portion of mashed potatoes. Have children create "land features," including mountains, valleys, and mesas.
2. Have children predict what will happen when gravy, representing rainfall, is poured on.
3. Pour gravy on potatoes and observe results.
4. Children can then eat their creations!

■ Gelatin Creations

NOTE: Here the gelatin represents the earth. The dinosaur-shaped fruit snacks represent fossils. The children are paleontologists.

Materials

8-ounce package of flavored gelatin
2 cups water
dinosaur-shaped fruit snacks (available in grocery stores)
mixing bowl

mixing spoon
clear plastic cups
plastic spoons
stove or microwave
refrigerator

Procedure

1. Combine gelatin and water according to package directions.
2. Let partially set in mixing bowl.
3. Spoon partially set gelatin into clear plastic cups.
4. Stir in dinosaur-shaped fruit snacks.
5. Place cups in refrigerator and let gelatin set completely.
6. At serving time, children become paleontologists. After they discover the "fossils," they can eat the fruit snacks.

■ Spicy Fried Dough
[Makes about 2 1/2 cups]

NOTE: This project requires lots of effort, but it is worthwhile. Obviously these creations do not last long.

Materials

1 teaspoon cinnamon	3 tablespoons vegetable oil
1/2 teaspoon nutmeg	mixing bowl
2 cups all-purpose flour	mixing spoon
2 teaspoons baking powder	wax paper
1/3 cup sugar	frying oil
1/3 cup milk	deep fryer

Procedure

1. Combine cinnamon, nutmeg, flour, baking powder, and sugar in mixing bowl.
2. Add milk and 3 tablespoons vegetable oil.
3. Knead until dough forms a ball.
4. Divide into portions and place on pieces of wax paper.
5. Mold into desired shapes.
6. Fry in oil at 375°F until brown.
7. Eat!

Gingerbread Houses

Gingerbread can be dated back to the eleventh century. The ginger helped preserve the finished products. Children enjoy making gingerbread houses, but a gingerbread house takes a great deal of time.

- First make "Gingerbread house templates" (below).
- Next prepare "Gingerbread Dough" (see p. 319).
- Prepare "Royal Icing" (see p. 317).
- Prepare "Sugar Glass" (see p. 318).
- Follow directions for "Constructing the Gingerbread House" (see p. 320).

■ Gingerbread House Templates

NOTE: The templates make a very simple, basic house. As children become more expert, they can make churches, stores, or whatever strikes their imagination.

Procedure

1. Make wax paper templates as follows:
2. Cut 2 pieces wax paper, 7 1/2 by 4 1/4 inches. One piece will be the front, and the other piece will be the back. Cut windows and doors in wax paper as desired.
3. Cut 2 pieces wax paper for sides of house: each piece should have 5 edges and be shaped for a house with a peak. The rectangular bottom of each piece should be 7 by 4 1/4 inches. The triangular portion should be 7 inches wide and 4 1/4 inches high. Cut windows in wax paper as desired.
4. Cut 2 pieces wax paper 9 by 6 1/4 inches to serve as roof sections.

■ Royal Icing
[Makes enough for 1 house]

NOTE: Royal icing is not tasty. Because royal icing dries very hard, it serves as mortar, joining gingerbread pieces together.

Materials

3 egg whites	mixing bowl with electric mixer
1/2 teaspoon cream of tartar	mixing spoon
1 pound confectioners' sugar	refrigerator

Procedure

1. Combine ingredients in mixing bowl.
2. Beat at high speed until somewhat stiff peaks form.
3. Start gluing!
4. Unused icing can be refrigerated for 2 days in a container with a tight lid. However, old icing will take longer to dry. Make fresh royal icing for any detail work.

■ Sugar Glass
[Makes plenty of sugar glass for gingerbread house]

Materials

2 cups sugar
2/3 cup white corn syrup
1/3 cup water
pot
stove or heating element
mixing spoon

candy thermometer
food coloring
1/2 teaspoon vanilla extract (optional)
cookie sheets with lip
nonstick cooking spray

Procedure

1. Combine sugar, white corn syrup, and water in pot.
2. Bring to a boil and cook, stirring constantly, until candy thermometer reads 300°F.
3. Add food coloring and vanilla extract if desired.
4. Spray cookie sheets with nonstick cooking spray.
5. Pour extremely hot sugar solution onto cookie sheets.
6. Allow sugar glass to harden.
7. Break sugar glass into pieces and insert pieces into window frames in gingerbread houses.

■ Gingerbread Dough

[Makes enough for 1 house]

Materials

6 cups all-purpose flour	2 large eggs
1 teaspoon ground ginger	nonstick cooking spray
1 teaspoon ground cloves	2 large mixing bowls
1 1/2 teaspoons ground nutmeg	mixing spoon
1 tablespoon ground cinnamon	floured surface for kneading
3/4 teaspoon salt	cookie sheets
1 teaspoon baking soda	rolling pin
1/2 cup (1 stick) butter at room temperature	sharp knife
	paper towels
1/2 cup shortening	refrigerator
1 cup sugar	oven
1 1/2 cups molasses	

Procedure

1. Combine 5 cups flour with spices, salt, and baking soda in a mixing bowl.

2. Mix butter and shortening in other mixing bowl. Beat in sugar.

3. Add molasses and eggs to butter–shortening mix.

4. Slowly add dry ingredients to butter–shortening mix.

5. Turn dough onto floured surface and knead, adding more flour if necessary to make a stiff dough.

6. Let dough rest for 1 hour so that it rolls out well. Refrigerate for the last half hour.

7. Spray cookie sheets with nonstick cooking spray.

8. Preheat oven to 300°F.

9. While oven is preheating, organize gingerbread house templates.

10. Place dough on cookie sheets and roll with rolling pin. The dough should be 1/4 inch thick. Lay wax paper gingerbread house templates over rolled-out dough and cut shape with sharp knife.

11. Remove wax paper.

12. Bake thin gingerbread for 6 minutes. Thicker gingerbread may need to bake between 10 and 15 minutes.

13. Remove baked pieces of gingerbread and place on paper towels to dry out overnight.

14. Any dough not baked can be wrapped and stored in refrigerator. Remove from refrigerator 30 minutes before rolling and baking.

■ Constructing the Gingerbread House

1. Create a rigid base for the house. Metal trays, pizza circles, and plastic platters work well.

2. Cut a piece of wax paper 7 by 7 1/2 inches to mark the floor. Attach it to base with a bit of royal icing.

3. Apply royal icing along bottom edge of gingerbread house front.

4. Align baked house front with front edge of wax paper floor and place it on base. It will need to be propped up temporarily.

5. Apply royal icing along bottom edge of 1 side of house.

6. Align house side with side edge of wax paper floor and place it on base. Using royal icing, attach house side to house front.

7. Apply royal icing along bottom edge of gingerbread house back.

8. Align house back with back edge of wax paper floor and place it on the base. Using royal icing, attach house back and house side together.

9. Apply royal icing along bottom edge of other house side.

10. Align house side with side edge of wax paper floor and place it on base. Using royal icing, attach house side to house front. Then attach house side to house back.

11. Let house frame set for at least 1 hour.

12. Apply royal icing to top edges of all 4 pieces of gingerbread house frame.

13. Place roof pieces on the house frame. Using royal icing, join the 2 roof pieces.

14. Let entire gingerbread house dry for at least 1 hour before decorating.

15. Now comes the best part—decorating. The following are some gingerbread house decorating ideas:

   - Inverted ice cream cones with pointed ends make nice pine trees.
   - Small pretzel sticks become excellent fencing.
   - Vanilla wafers make great shingles.
   - Pretzel rods can become chimneys or even walls of log cabins.
   - Animal crackers can provide some interesting gingerbread pets.
   - Pieces of licorice whip can divide sections of walls into bricks.
   - Tinted coconut simulates grass.
   - Granola can become gravel.
   - Colored sugars make everything glitter.
   - Marshmallows can become smoke puffs from a chimney.

■ Easy Gingerbread Houses
[Makes 1 house]

NOTE: Easy gingerbread houses may not be as elaborate as those made from scratch. However, they take less time and are particularly successful with younger children.

Materials

6 double graham crackers
small, serrated knife
Royal Icing (see p. 317).
decorating materials such as licorice strings, pretzels, candies, or vanilla wafers

base such as a metal tray, pizza circle, or plastic platter
props such as bottles of glue or containers of paint

Procedure

1. Two of the double graham crackers will be front and back walls, 2 will be the sides, and 2 will be the roof. To make peaks at the tops of the walls, cut with the knife using a sawing motion.
2. Spread bottom edge of front wall with royal icing and attach to base. Prop with bottles of glue until icing sets.
3. Spread royal icing on the sides and bottom of a side wall (1 long edge and 2 short edges). Attach side wall to front wall and to base. Prop with bottles of glue until icing sets.
4. Spread side and bottom edges of back wall with royal icing and attach to standing side and base. Prop with bottles of glue until icing sets.
5. Spread other side wall graham cracker with royal icing on the 2 sides and bottom edges. Attach to base, front wall, and back wall. Prop with bottles of glue until icing sets.
6. Allow to set and remove any props.
7. The last 2 graham crackers will form the roof. Join the 2 crackers' long sides together with royal icing.
8. Apply royal icing to the top edges of the front and back walls.
9. Place roof graham crackers on the wall graham crackers, forming a peak. Let set.
10. Decorate and display.

28
Dairy Products

Making butter, cheese, and other dairy products are all excellent hands-on projects for children. Because these foods date back thousands of years, they relate well to studies of ancient civilizations as well as the American colonial period. These projects also can promote healthy eating habits by relating a basic food, milk, to the processes of science.

All containers and utensils used to make milk products must be exceptionally clean.

■ Sweet Butter
[Makes about 1 cup]

NOTE: Making butter is a good way to end a study of the dairy region. Another way to make butter is to pour the cream into a large bowl and beat with an electric mixer until butter separates from buttermilk. The following way is much more fun!

Materials

1 pint heavy cream	colander
4 small extremely clean containers with tight-fitting lids	large bowl
	large spoon
4 extremely clean marbles	salt (optional)

Procedure

1. Pour heavy cream into containers.

2. Place an extremely clean marble in each container and add tight-fitting lids.

3. Begin shaking the containers and shake and shake and shake. Pass container among friends.

4. At some point the children can no longer hear the marble in the jar when the jar is shaken. At this point a type of unsweetened whipped cream has formed.

5. Continue to shake and shake and shake. Soon the children can hear the marble again inside the jar. The butter has formed.

6. Place colander inside large bowl.

7. Remove tight-fitting lids and pour contents into colander. Strain mixture. The solids are butter, and the liquid is buttermilk.

8. Press butter with spoon to remove even more buttermilk.

9. Add salt if desired. Children can now taste their homemade butter!

■ Cinnamon–Raisin Butter
[Makes about 3/4 cup]

NOTE: Top Irish soda bread with this easy-to-make butter.

Materials

1/2 cup (1 stick) softened butter
1 teaspoon cinnamon
3 tablespoons confectioners' sugar

1/4 cup raisins
mixing bowl
mixing spoon

Procedure

1. Blend together butter, cinnamon, and confectioners' sugar.

2. Gently fold in raisins and serve.

■ Maple–Nut Butter
[Makes about 3/4 cup]

NOTE: Lavish this butter on hot corn bread (see p. 289). The corn bread immediately soaks up the butter.

Materials

1/2 cup (1 stick) softened butter
1/4 cup maple syrup
1 tablespoon chopped pecans

mixing bowl
mixing spoon

Procedure

1. Combine ingredients in mixing bowl and serve.

■ Honey Butter
[Makes about 3/4 cup]

NOTE: Honey butter and biscuits (see p. 286) make a tasty combination.

Materials

1/2 cup (1 stick) softened butter
1/4 cup honey

mixing bowl
mixing spoon

Procedure

1. Combine ingredients in mixing bowl and serve.

■ Orange Butter
[Makes about 1/2 cup]

NOTE: For a tasty treat, spread orange butter on orange–honey bread (see p. 285) and top with orange–pineapple marmalade (see p. 373).

Materials

1/2 cup (1 stick) softened butter   mixing bowl
1 1/2 teaspoons grated orange zest   mixing spoon
3 tablespoons confectioners' sugar

Procedure

1. Combine ingredients in mixing bowl and serve.

■ Whipped Cream
[Makes 2 cups]

NOTE: Children could use a mixer instead of the jars, but shaking is much more fun.

Materials

1 pint heavy cream   4 extremely clean marbles
3 tablespoons sugar   refrigerator
1 teaspoon vanilla extract
4 small extremely clean containers
    with tight-fitting lids

Procedure

1. Put containers and marbles in refrigerator for about a half hour before beginning project. Cold containers and cold cream bring best results.
2. Remove containers from refrigerator. Pour heavy cream into containers.
3. Add sugar and vanilla.
4. Place an extremely clean marble in each container and add tight-fitting lids.
5. Begin shaking the containers and shake and shake and shake. Pass container among friends.
6. At some point the children can no longer hear the marble in the jar when the jar is shaken. Whipped cream is ready to taste!

■ Ice Cream without an Ice Cream Maker

[Makes about 2 3/4 cups ice cream—enough for 6 children]

NOTE: This ice cream requires no expensive machinery. Many children can participate in the process.

Materials

1 rennet tablet (available in the pudding section of grocery stores; also called junket)
1 tablespoon cold water
1/3 cup sugar
1 1/2 cups milk
1 cup heavy cream
2 teaspoons vanilla

small bowl
stainless steel pot
mixing spoon
freezer-safe container
large fork
electric mixer
stove or heating element
refrigerator with freezer

Procedure

1. In small bowl, dissolve rennet tablet in 1 tablespoon cold water.

2. Combine sugar, milk, and heavy cream in stainless steel pot. Heat, stirring constantly, until mixture is lukewarm.

3. Add vanilla and rennet.

4. Pour mixture into freezer-proof container and let stand 15 minutes.

5. Place container in freezer until mixture is fairly frozen but not completely hard (about 1 hour).

6. Remove container from freezer. Use fork to break mixture into large pieces.

7. Whip mixture with electric mixer for 3 minutes.

8. Return mixture to freezer until completely frozen.

■ Curds and Whey
[Makes 1 quart]

NOTE: Water comprises about 88 percent of milk's content. The remaining 12 percent is composed of milk solids, a combination of caseins (proteins), milk sugar, milk fats, vitamins, and minerals. Curds and whey result when vinegar is added to milk. The vinegar curdles the caseins, producing curds. The resulting liquid is whey. Children can pretend to be Little Miss Muffet, but they should not eat the curds and whey like she did. They can eat cottage cheese (see recipe below) made from curds and whey.

Materials

1 quart whole milk   container
1/2 cup white vinegar   spoon

Procedure

1. Pour milk into container. Add white vinegar and stir.
2. Let mixture sit for 2–3 minutes. The milk will separate into curds (solid portion) and whey (liquid portion).

■ Cottage Cheese
[Makes 1 cup—enough for 3 children]

NOTE: Start draining curds early in the day so that cottage cheese can be eaten the same day.

Materials

curds from "Curds and Whey" (see   cheesecloth
    previous recipe)   salt
colander

Procedure

1. Line colander with cheesecloth. Pour curds into cheesecloth.
2. Cover with another piece of cheesecloth. Let drain.
3. Cut into smaller pieces and add salt to taste.
4. Cottage cheese is ready to eat!

■ Yogurt
[Makes 1/2 gallon—enough for 12 children]

NOTE: This plain yogurt has a tart, tangy taste. Children enjoy watching a little bit of yogurt become 1/2 gallon!

Materials

1/2 gallon milk
1 cup plain yogurt with live cultures (without additives or preservatives)
large stainless steel pot with lid
spoon

small, extremely clean containers with lids
large towel
fruit (optional)
stove or heating element
refrigerator

Procedure

1. Heat milk over medium heat in stainless steel pot until milk almost boils. Simmer for 15 minutes. This will kill any bacteria. Cool until lukewarm.
2. Stir in yogurt. Pour yogurt–milk mixture into containers and add lids.
3. Clean pot. Put containers of yogurt into pot.
4. Fill pot nearly up to lids with hot tap water. Place lid on pot. Wrap towel around pot.
5. Let stand for 5 hours. Open 1 container. The contents will be thickened to the consistency of yogurt.
6. Remove containers from pot and refrigerate. Add fruit if desired at serving time.

■ Yogurt Cheese
[Makes 1 cup]

NOTE: Yogurt cheese resembles cream cheese. It can be used to make dips.

Materials

3 cups plain yogurt
colander

cheesecloth
container with lid

Procedure

1. Place colander in sink and line with cheesecloth.
2. Empty yogurt into cheesecloth.
3. Allow yogurt to drain for about 18 hours. You now have yogurt cheese.
4. Refrigerate yogurt cheese in container with lid.

■ Frozen Yogurt Treats

[Makes 4 3/4 cups—enough for 10 children]

NOTE: Strawberries and bananas make a good fruit combination. So do blueberries and peaches.

Materials

2 cups fresh fruit

1/4 cup juice concentrate

2 1/2 cups plain yogurt

2 teaspoons vanilla

blender

10 paper cups

10 craft sticks

refrigerator with freezer

Procedure

1. Puree fruits and juice in blender.

2. Mix in yogurt and vanilla.

3. Pour mixture into paper cups. Place in freezer.

4. When mixture is fairly stiff, add craft sticks to serve as handles. Freeze at least 1 more hour.

5. Remove treats from freezer and peel off paper cups.

■ Greek Yogurt
[Makes about 4 cups]

NOTE: Greek yogurt is thicker than regular yogurt because the former contains less whey.

Materials

1/2 gallon milk
1 tablespoon Greek yogurt with live cultures
pot
stove or heating element
candy thermometer
large oven-proof bowl

mixing spoon
plastic wrap
large towel
cheesecloth
sieve
large bowl

Procedure

1. Pour milk into pot. Heat milk to a temperature of 175°F.

2. Remove milk from heat and pour it into large oven-proof bowl. Cool until it reaches a temperature of 115°F.

3. Stir in Greek yogurt and cover with plastic wrap. Wrap large towel around bowl and place in an oven.

4. Place oven-proof bowl in oven. Turn on the oven to 100°F. Turn off oven when temperature reaches 100°F. Keep oven door open so that oven light remains on and temperature remains around 100°F.

5. Keep Greek yogurt in oven for 8–12 hours until yogurt is thick.

6. Line sieve with cheesecloth and place sieve in large bowl.

7. Pour Greek yogurt into cheesecloth so that whey can drip into bowl for about an hour.

8. Refrigerate Greek yogurt.

■ Cream Cheese
[Makes 1 cup]

NOTE: Flavored cream cheese spreads have become quite popular. For example, stir in raisins and a bit of cinnamon to top a toasted bagel.

Materials

3 cups heavy cream
4 tablespoons buttermilk
mixing bowl

mixing spoon
colander
splatter screen

Procedure

1. Combine heavy cream and buttermilk in mixing bowl.
2. Line colander with cheesecloth and place in sink.
3. Pour cream–buttermilk mixture into cheesecloth.
4. Wash and dry mixing bowl.
5. Place splatter screen over mixing bowl.
6. Place cheesecloth on screen so that liquid will drain into bowl.
7. Leave for at least a day. The longer the cheese stands, the drier it will be.

■ Hard Cheese
[Makes about 2 pounds]

NOTE: This basic recipe makes a cheese somewhat like a Colby. It can be modified by changing the type of milk, varying the cooking time, or introducing cultures other than buttermilk. However, this recipe allows children to be part of all the steps. The actual cheese making takes only a day, but pressing the cheese adds another 5 days. The cheese should mellow for at least a month before it can be consumed. This activity would be a great way to salute the state of Wisconsin, "America's Dairyland."

Materials

1 gallon pasteurized milk
1 cup cultured buttermilk
1/8 rennet tablet (available in the pudding section of grocery stores; also called junket)
1/4 cup cool water
3 tablespoons coarse, noniodized salt
small amount of vegetable oil
2 stainless steel pans, 1 large enough to contain the gallon of milk but small enough to fit inside other
pan (together the pans act as a double boiler)
small bowl
small mixing spoon
knife
candy thermometer
large mixing spoon
cheesecloth
colander
cheese press (see process below)
stove or heating element

Procedure

1. Place smaller stainless steel pot into larger one. Add milk to smaller pot.
2. Fill larger pot with water until water level is equal to milk level.
3. Heat milk to 88°F. Remove pot from heat and add cultured buttermilk.
4. Allow mixture to rest for at least 1/2 hour but up to 3 hours. The longer it rests, the sharper the cheese will be.
5. Place rennet tablet in small bowl and add water to dissolve.
6. Pour rennet mixture into milk mixture and stir. Let sit for several hours until milk has coagulated into curd.
7. Using knife, cut curd into very small pieces.
8. Knead curd for 10 minutes, until whey separates.
9. Gradually heat until curds (still inside double boiler) reach 102°F. Maintain that temperature for 1 hour and stir frequently.
10. Line colander with cheesecloth. Pour in curds. Strain whey away.
11. Slowly stir in salt.

12. Remove curd-filled cheesecloth from colander and tie a knot in cheesecloth.

13. Place wrapped curds in cheese press. Apply a bit of pressure with 1 brick for 1 hour.

14. Increase pressure by using both bricks. Press for 5 hours.

15. Remove cheese from press. Let dry in a cool-well-ventilated area for 5 days.

16. Coat cheese with vegetable oil to prevent molding. Return cheese to a cool, well-ventilated area. Let cheese ripen for 1 month.

■ Cheese Press

NOTE: This homemade cheese press has 2 purposes. It gives the cheese a shape, and it forces liquids out of the developing cheese.

Materials

2 bricks	sink
2 plastic bags	splatter screen
dinner plate	springform pan with bottom

Procedure

1. Place a brick into each plastic bag and secure. These bricks will be weights.

2. Invert dinner plate and place it on sink bottom.

3. Place splatter screen on top of plate.

4. Open springform pan and remove bottom. Do not secure clamp of the side of springform pan.

5. Place springform pan on top of splatter screen.

6. Place wrapped cheese inside springform pan.

7. Place springform pan bottom on top of cheese.

8. Align bricks on top of springform pan bottom. The springform pan bottom will distribute the weight of the bricks.

CHAPTER

29
Treats Children Can Make

Children love to make candy and other treats. Here are plenty of easy-to-use recipes. Many of these recipes require little time because they use a microwave oven instead of a stove. Only 1 recipe needs a candy thermometer.

■ Healthy S'Mores
[Makes 5 treats]

NOTE: Graham crackers and peanut butter are reasonably healthy. Other nut butters could be substituted for peanut butter. Other fruits, such as sliced strawberries, could be used.

Materials

10 graham crackers	1 banana
1/2 cup creamy peanut butter	butter knife

Procedure

1. Spread peanut butter on 5 graham crackers.
2. Slice banana and place slices on top of peanut butter.
3. Top banana slices with remaining graham crackers to make sandwiches.

■ Gorp
[Makes 3 cups—enough for 6 children]

NOTE: Gorp stands for Gold Old Raisins and Peanuts. This recipe adds some sweetness as well. Gorp makes a good mid-morning snack. It is also great for carrying on hikes.

Materials

1 cup raisins	mixing bowl
1 cup M&M® candies	mixing spoon
1/2 cup salted peanuts	small snack bags
1/2 cup unsalted peanuts	

Procedure

1. Mix all ingredients in bowl and then disperse in snack bags.

■ Caramel Corn
[Makes 2 1/2 quarts]

NOTE: Melting the caramels in the microwave makes this recipe fun.

Materials

28 unwrapped caramels
2 tablespoons cold water
2 1/2 quarts popped popcorn
microwave-safe bowl
mixing spoon

large mixing bowl
cookie sheet
nonstick cooking spray
microwave oven
oven

Procedure

1. Place caramels and water in microwave-safe bowl.

2. Microwave caramels and water at high setting for 1 1/2 minutes.

3. Stir. If caramels are not melted, microwave at high for 30 more second intervals until melted.

4. Place popcorn into large mixing bowl. Pour caramel sauce over popcorn and stir until popcorn is coated.

5. Spray cookie sheet with nonstick cooking spray.

6. Pour popcorn-caramel mixture onto cookie sheet.

7. Spread mixture evenly over cookie sheet.

8. Bake at 250°F for 20 minutes. Break into pieces.

■ Cheese Popcorn
[Makes 2 quarts]

NOTE: Plain popcorn is very nutritious. The cheese adds extra flavor.

Materials

1/4 cup (1/2 stick) butter, melted
1/2 cup grated Parmesan cheese
2 quarts popped popcorn

small bowl
large bowl
mixing spoon

Procedure

1. Mix butter and cheese in small bowl.

2. Pour popcorn into large bowl.

3. Pour butter–cheese mixtures over popcorn. Stir to evenly coat popcorn.

■ Popcorn Balls
[Makes 12 balls, each 3 inches in diameter]

NOTE: The sugar mixture can be very hot, so children should be very careful.

Materials

1/2 cup sugar
1/2 cup brown sugar
1/2 cup light corn syrup
1/3 cup water
1 cup (2 sticks) butter, cut into pieces
4 cups popped popcorn
enough butter to butter hands
pot

stove or heating element
candy thermometer
large mixing bowl
spoon
nonstick cooking spray
wax paper
plastic wrap

Procedure

1. Combine sugars and corn syrup in pot and bring to a boil.
2. Add butter and stir. Cook for 20–30 minutes, stirring constantly, until the mixture reads 300°F on candy thermometer.
3. Coat large mixing bowl with nonstick cooking spray. Pour popcorn into large mixing bowl.
4. Drizzle syrup over popcorn and mix.
5. Butter hands so that the popcorn will not stick to skin.
6. As soon as mixture cools, quickly shape into 3-inch balls.
7. Place balls on wax paper until they cool.
8. Wrap balls individually in plastic wrap to keep them fresh.

■ Crisp Rice Treats
[Makes 15 treats]

NOTE: Recipes for crisp rice treats have been around for a long time. This recipe uses the microwave oven to speed the process. The recipe can be modified by adding 1/4 cup peanut butter to the marshmallow–butter mixture. Raisins, dried fruit, or chocolate chips can be added before the mixture is placed in the pan.

Materials

1 package (10 ounces) marshmallows
    or 4 cups mini-marshmallows
1/4 cup (1/2 stick) butter
6 cups crisp rice cereal
microwave-safe bowl
mixing spoon

nonstick cooking spray
pan, 13-inches by 9-inches by
    2-inches
spatula
microwave oven

Procedure

1. Combine butter and marshmallows in microwave-safe bowl. Microwave at high setting for 2 minutes.
2. Stir and microwave at high setting for 1 more minute.
3. Add cereal and stir until cereal is coated with butter–marshmallow mixture.
4. Coat pan and spatula with nonstick cooking spray.
5. Pour cereal–marshmallow mixture into pan.
6. Evenly distribute mixture with coated spatula.
7. Cool and cut into squares.

■ Jelly Bean Nests
[Makes 12 nests]

NOTE: These treats are great to make around Easter.

Materials

2 cups miniature marshmallows
1/4 cup (1/2 stick) butter
4 cups chow mein noodles
extra butter to coat hands and lubri-
  cate muffin pan

jelly beans
microwave-safe mixing bowl
microwave oven
spoon
12-cup muffin pan

Procedure

1. Combine marshmallows and butter in microwave-safe mixing bowl.

2. Microwave on high for about 3 minutes, stirring every minute or so, until the marshmallows are melted.

3. Stir in chow mein noodles.

4. Butter muffin cups.

5. With buttered fingers press some of the mixture into each of the muffin cups.

6. Refrigerate for several hours.

7. Remove nests from muffin cups and add jelly beans.

■ Homemade Marshmallows
[Makes about 20]

NOTE: The gelatin forms a colloid, and air is trapped within the colloid to produce puffy marshmallows. Children could make colored marshmallows by adding a few drops of food coloring at step 3.

Materials

2 tablespoons (2 envelopes) unflavored gelatin
1 cup boiling water
1 cup sugar
2 1/2 teaspoons vanilla
confectioners' sugar
mixing bowl

mixing spoon
egg beater
wax paper
nonstick cooking spray
shallow pan
knife

Procedure

1. Combine unflavored gelatin and boiling water in mixing bowl.

2. Gradually add sugar and vanilla.

3. Beat with egg beater until mixture becomes marshmallow cream. This may take about 15–20 minutes.

4. Spray wax paper with nonstick cooking spray and place in a shallow pan.

5. Pour marshmallow cream onto wax paper.

6. Cover loosely with more wax paper. Make sure top wax paper does not touch marshmallow cream.

7. Let pan sit overnight. The next day cut hardened marshmallow mixture into cubes and dust with confectioners' sugar.

■ Marshmallow Delights
[Makes 40]

NOTE: Not many snacks are this easy to make. Children can set up an assembly line of "dippers" and "toppers." Foods other than marshmallows could also be dipped. Possibilities include pieces of fruit, cake cubes, and cookies.

Materials

1 package (10 ounces) marshmallows
1 package (12 ounces) semisweet
    chocolate chips
small containers of shredded coconut,
    sprinkles, and chopped nuts
large, flat pan

microwave-safe bowl
fork
wax paper
microwave oven
refrigerator and freezer

Procedure

1. Lay marshmallows on large, flat pan.
2. Place pan in freezer for 15 minutes.
3. Pour semisweet chocolate chips into microwave-safe bowl and microwave at high setting for about 2 minutes or until chips are melted.
4. Using the fork, dip frozen marshmallows into melted chocolate.
5. Roll chocolate-coated marshmallows in toppings and place on wax paper.
6. Chill in refrigerator and eat.

■ Chocolate Marshmallow Spiders
[Makes 24 treats]

NOTE: These treats are great to make at Halloween.

Materials

8 squares semisweet baking chocolate   microwave oven
2 cups miniature marshmallows   mixing spoon
string licorice   wax paper
various candies   tray
microwave-safe bowl   refrigerator

Procedure

1. Place baking chocolate in microwave-safe bowl and microwave until melted (about 2 minutes).

2. Stir until completely melted.

3. Add marshmallows and stir until combined with the chocolate.

4. Line tray with wax paper.

5. Drop mixture by spoonfuls onto wax paper. Each spoonful is now a "spider's body."

6. Add licorice strings to form legs.

7. Use candies to create eyes and other details.

8. Refrigerate until firm.

■ Marshmallow Ghosts
[Makes about 30 ghosts]

NOTE: These goodies are also fun to make at Halloween.

Materials

4 packages (6 ounces each) white
   chocolate baking squares
3 cups miniature marshmallows
various candies or icing
microwave-safe bowl

microwave oven
mixing spoon
wax paper
trays

Procedure

1. Microwave white chocolate baking squares in microwave-safe bowl until melted (about 3–4 minutes).
2. Stir to make mixture smooth. Let cool for a few minutes.
3. Add marshmallows and combine.
4. Place wax paper on trays.
5. Drop mixture by spoonfuls on wax paper. Each spoonful is now a marshmallow "ghost."
6. Decorate with candies and icing.

■ Pemmican
[Makes about 3 cups, enough for at least 12 children]

NOTE: The original Native American fast food, pemmican uses buffalo fat instead of peanut butter. However, peanut butter is healthier, easier to find, and more appealing than animal fat.

Materials

1 cup jerky
1 cup dried berries
1 cup roasted nuts
2 teaspoons honey
1/4 cup peanut butter
other spices such as garlic powder or
   paprika to taste

food processor
mixing bowl
mixing spoon
small paper cups
disposable spoons

Procedure

1. Use food processor to grind jerky into a powder.

2. Add dried berries and nuts.

3. Add honey, peanut butter, and spices. Combine.

4. Distribute mixture into cups with spoons so that children can each have a taste.

■ Homemade Peanut Butter
[Makes 2 cups]

NOTE: Homemade peanut butter may separate after a while. Simply stir to blend the oil back in. Store in the refrigerator.

Materials

3 cups shelled, roasted peanuts
small amount of vegetable oil

blender

Procedure

1. Pour peanuts into blender and pulse until smooth.

2. Add a bit of oil if necessary to increase smoothness.

■ Chocolate Peanut Butter Treats
[Makes 15 treats]

NOTE: These no-bake treats are sweet, salty, and crunchy all at the same time.

Materials

1 package (6 ounces) semisweet chocolate chips

2/3 cup chunky peanut butter

1 container (3 ounces) chow mein noodles

microwave-safe bowl

mixing spoon

aluminum foil

microwave oven

Procedure

1. Combine peanut butter and semisweet chocolate chips in microwave-safe bowl.

2. Microwave at high setting for 3 minutes or until melted.

3. Add chow mein noodles and stir.

4. Drop by spoonfuls onto aluminum foil and let cool.

■ Chocolate Peanut Butter Pizza
[Makes 16 small slices]

NOTE: Children can learn about fractions as they cut their chocolate peanut butter pizzas. Make sure no student is allergic to nuts.

Materials

1 6-ounce package semisweet chocolate chips
1 6-ounce package peanut butter chips
2 ounces white baking chocolate
various candies
2 microwave-safe bowls

microwave oven
mixing spoon
1 12-inch pizza pan
nonstick cooking spray
spatula
pizza wheel or knife

Procedure

1. Microwave semisweet chocolate chips and peanut butter chips in 1 microwave-safe bowl until melted (about 2 minutes).
2. Stir to blend the chips.
3. Spray pizza pan with nonstick cooking spray.
4. Spread melted mixture onto the pizza pan with spatula.
5. Microwave white baking chocolate in the other microwave-safe bowl until it is melted (about 1 minute).
6. Drizzle mixture over melted chocolate–peanut butter chip mixture so that it looks like cheese.
7. Decorate with various candies.
8. Refrigerate for about a half hour.
9. Cut into 16 wedges and serve.

■ Peanut Butter Turtles

[Makes 10–12]

NOTE: Make sure no one is allergic to nuts because this recipe uses peanut butter. Children could learn about real turtles as they munch their peanut butter turtles.

Materials

1/2 cup peanut butter
3 ounces cream cheese, softened
2 tablespoons honey
1 cup granola or crunchy cereal
slivered almonds
raisins

mixing bowl
mixing spoon
wax paper
tray
refrigerator

Procedure

1. Combine peanut butter, cream cheese, and honey in mixing bowl.
2. Gently mix in cereal.
3. Place wax paper on tray.
4. Drop mixture by spoonfuls on tray.
5. Shape each mixture to look like the body of the turtle.
6. Place slivered almonds to look like legs and the raisins to look like eyes.
7. Refrigerate until firm.

■ Peanut Butter Fudge

[Makes about 20 pieces]

NOTE: This fudge tastes good, but it remains somewhat soft when it is at room temperature.

Materials

1 18-ounce jar of peanut butter
1 can vanilla frosting
mixing bowl
mixing spoon

1 baking pan, 9 inches by 9 inches
refrigerator
airtight container

Procedure

1. Combine peanut butter and frosting in mixing bowl.
2. Pour mixture into the pan.
3. Refrigerate.
4. Store in an airtight container.

■ Peanut Brittle
[Makes 3 cups—enough for 6 children]

NOTE: Peanut brittle stores well for a long time. The peanuts add some nutrition to this candy.

Materials

3 tablespoons butter
1 cup peanuts
3 cups sugar
dinner plate
aluminum foil

saucepan
mixing spoon
stove or heating element
hot pads

Procedure

1. Cover dinner plate with aluminum foil. Generously grease foil with butter.

2. Distribute peanuts evenly on plate.

3. Heat sugar in saucepan at high heat. When sugar begins to melt, turn down heat and stir constantly.

4. When sugar is completely melted, remove saucepan from heat with hot pads. Pour melted sugar over peanuts.

5. Let cool. Peel peanut brittle from aluminum foil and break into pieces.

■ Cereal Snacks
[Makes 9 cups]

NOTE: The peanut butter and cereal make this snack somewhat nutritious. It stores well.

Materials

3/4 cup semisweet chocolate chips

1/2 cup peanut butter

1/2 cup butter

1/2 teaspoon vanilla

9 cups nonflake cereal (e.g., Corn Chex®)

1 1/2 cups confectioners' sugar

microwave-safe bowl

large bowl

mixing spoon

gallon-size plastic storage bag

wax paper

small sandwich bags

microwave oven

Procedure

1. In microwave-safe bowl, combine semisweet chocolate chips, peanut butter, and butter.

2. Microwave at high setting for 1 minute or until semi-sweet chocolate chips have melted

3. Add vanilla.

4. Pour cereal into large bowl.

5. Pour chocolate chip–peanut butter mixture over cereal, stirring to coat cereal pieces.

6. Pour mixture into gallon-size plastic storage bag. Add confectioners' sugar and close bag. Shake bag until cereal is well coated.

7. Pour cereal snacks onto wax paper and let cool.

8. Pour mixture into small sandwich bags and distribute.

■ Pizza Muffins
[Each English muffin half makes an individual pizza]

NOTE: These quick and filling pizza muffins are a great finish to a unit on nutrition. Each muffin uses all four basic food groups.

Materials

English muffins, split
pizza sauce
grated mozzarella cheese
diced green peppers, mushrooms,
    olives, onions, pepperoni

spoon
cookie sheet
oven broiler or toaster oven

Procedure

1. Spread a spoonful of pizza sauce onto each English muffin half.

2. Sprinkle on cheese.

3. Add toppings to taste.

4. Place muffin halves onto cookie sheet and broil in oven until cheese melts.

■ Latkes
[Makes 15 latkes]

NOTE: Latkes are Jewish potato pancakes. They are served alone or topped with applesauce or sour cream. While latkes are delicious anytime, they are part of traditional meals at Hanukkah.

Materials

5 large potatoes

1 small onion

2 large eggs

1/2 teaspoon salt

3/4 cup flour

1/2 teaspoon baking powder

shortening to grease electric frying pan

wax paper

potato peeler

grater

mixing bowl

mixing spoon

electric frying pan

spatula

applesauce or sour cream (optional)

serving plate

Procedure

1. Cover work area with wax paper.
2. Peel potatoes and save peels for bird food or for vegetable stock.
3. Beat eggs in mixing bowl.
4. Grate potatoes and add them to mixing bowl.
5. Chop onions and add to mixing bowl.
6. Add salt, flour, and baking powder.
7. Heat electric frying pan to a high temperature. Grease with shortening.
8. Drop spoonfuls of potato mixture into frying pan. Let brown. Turn latkes with spatula and let brown on other side.
9. Place browned latkes on serving plate and continue to prepare the rest of the latkes.
10. Serve with apple sauce or sour cream if desired.

■ Matzo
[Makes 8–10 matzo, enough to serve a group of about 30 children]

NOTE: Matzo, an unleavened bread, is eaten by Jews during Passover. Jews do not eat leavened bread during Passover.

Materials

2 cups matzo meal
about 1 1/2 cups water
3/4 teaspoon salt
mixing bowl
mixing spoon
kneading surface with extra matzo meal

wax paper
rolling pin
fork
cookie sheets
oven
spatula

Procedure

1. In mixing bowl, combine matzo meal, salt, and enough water to make workable dough.
2. Turn dough onto kneading surface and knead. Cut dough into 8 pieces.
3. Place each piece between two pieces of wax paper and roll out to cracker thickness.
4. Prick surface of each with the fork.
5. Bake on ungreased cookie sheets at 475°F for about 4 minutes. Turn matzo over with the spatula and bake another 4 minutes until brown and crisp.

■ Matzo Balls

[Makes about 35 matzo balls]

NOTE: Matzo ball soup is served during Passover, but it is a favorite the rest of the year as well.

Materials

3 tablespoons chicken fat
3 large eggs
1 teaspoon salt
1/4 teaspoon pepper
1 cup matzo meal

mixing bowl
spoon
refrigerator
simmering chicken broth
stove

Procedure

1. Combine chicken fat and eggs in bowl.
2. Add salt and pepper.
3. Add enough matzo meal so that the mixture is slightly sticky.
4. Refrigerate for at least 2 hours.
5. With wet hands roll pieces of dough into balls about 3/4 inch in diameter.
6. Drop into simmering chicken soup and cook for 15 minutes.

■ Toasted Pumpkin Seeds

[Makes 2 cups]

NOTE: Children can save their seeds when they carve their Halloween pumpkins. Keep the toasted seeds until Thanksgiving and add them to the feast.

Materials

2 cups pumpkin seeds, cleaned and washed
1 tablespoon salt
water
bowl

nonstick cooking spray
cookie sheet
colander
extra salt (optional)
oven

Procedure

1. Pour seeds into bowl and cover with water.
2. Stir in salt and soak seeds for 4 hours.
3. Coat cookie sheets with nonstick cooking spray.
4. Drain seeds in colander and spread them on cookie sheet.
5. Shake on additional salt, if desired.
6. Bake at 300°F until seeds are amber brown, about 10–15 minutes.

■ Fruit Leather
[Makes enough for 20 children]

NOTE: Pioneers preserved their fall fruit harvest by making fruit leather. Today's prepackaged fruit-roll snacks are a modern version of fruit leather.

Materials

4 quarts fruit (such as apples, peaches, or apricots), peeled, cored, and cut
1–2 cups apple juice
honey
cinnamon
cornstarch

blender
large pot
cookie sheets
freezer paper
cheesecloth
cake rack
stove or heating element

Procedure

1. Puree fruit in blender.
2. Pour pureed fruit into large pot and add apple juice.
3. Over low heat, bring mixture to a boil.
4. Add honey and cinnamon according to taste.
5. Reduce temperature until mixture is just simmering. Simmer mixture, stirring often, until it has the consistency of apple butter.
6. Cover cookie sheets with freezer paper.
7. Pour mixture onto cookie sheets so that mixture forms a layer 1/4-inch thick.
8. Cover with cheesecloth and place in the sun to dry. Drying can take up to 10 days.
9. Fruit leather can also be dried in a low-temperature oven or food drier.
10. When fruit leather is dry enough to keep its shape, cut it into strips. Place strips on cake rack to make sure all sides dry.
11. Dry until the surface of the fruit leather is not longer sticky. Sprinkle cornstarch on strips and remove from cake rack. The cornstarch keeps the strips from clinging to one another.
12. Store in freezer paper.

■ Oven-Dried Fruit and Raisins
[Makes about half as much dried fruit as the initial amount of fresh fruit]

NOTE: Colonists dried apples by coring, slicing, and stringing them. The strings were placed near the fireplace to hasten the drying.

Materials

fresh fruit (see directions for preparation below)
cookie sheet

oven
storage containers

Procedure

1. Prepare fruit:

Preparing Grapes (Raisins)

   a. Wash white seedless grapes.

   b. Place them in boiling water until skins split.

   c. Drain.

Preparing Apples and Peaches

   a. Wash fruit.

   b. For peaches, peel skin and remove pits. For apples, peel and core.

   c. Cut into thin slices.

Preparing Berries and Cherries

   a. Remove cherry pits.

   b. Wash and drain.

2. Place fruit on cookie sheet.

3. Place cookie sheet in an oven at 150°F. Bake for 6 hours, or until fruit is dry but not brittle.

4. Cool and store in storage containers.

■ Sun-Dried Fruit and Raisins
[Makes about half as much dried fruit as the initial amount of fresh fruit]

NOTE: Remember to bring the fruit indoors each night. If the fruit is left out at night, the dew will plump up the fruit again.

Materials

fresh fruit (see directions for preparation of fruit in previous recipe)
large, flat pan

cheesecloth or netting
storage containers

Procedure

1. Prepare fruit and place in pan.
2. Cover pan with cheesecloth or netting to keep out bugs and birds.
3. Place pan in the sun. Turn fruit several times so that it will dry evenly.
4. At the end of the day, bring the pan inside.
5. The next day, put the pan back out in the sunshine.
6. Depending on temperature and humidity, the grapes should become raisins in 2 or 3 days. Larger pieces of fruit may take up to 5 days to dry.
7. Store fruit in storage containers in the refrigerator.

■ Natural Fruit Candy
[Makes 1 pound, about 20 pieces]

NOTE: Use only 1 type of dried fruit in the beginning. Otherwise, the colors may not be so pretty. Make sure no one is allergic to nuts.

Materials

1 pound dried fruit
about 1/2 cup fruit juice
1/2 cup finely chopped nuts
food processor

spoon
small bowl
wax paper
cookie sheet

Procedure

1. Pour dried fruit into food processor and grind.
2. Add enough fruit juice to make the ground fruit stick together.
3. Pour chopped nuts into the small bowl.
4. Line cookie sheet with wax paper.
5. Spoon out small amounts of fruit candy. Form small balls.
6. Coat small fruit balls in nuts and place on wax paper.
7. Store in an airtight container.

■ Irish Potato Candy

[Makes about 25]

NOTE: These candies are popular around St. Patrick's Day. Potatoes are an important part of Ireland's heritage, and these candies make St. Patrick's Day special. However, no real potatoes are used in this recipe.

Materials

1 8-ounce package of cream cheese	cinnamon
2 cups sifted confectioners' sugar	mixing bowl
1 ounce shredded coconut	spoon
1 teaspoon vanilla	wax paper

Procedure

1. Combine cream cheese, confectioners' sugar, and coconut in mixing bowl.
2. Spread cinnamon on wax paper.
3. Divide the mixture into portions about the size of a nickel.
4. Roll each portion into a ball and roll the portions in the cinnamon.
5. Keep refrigerated.

■ Irish Mashed Potato Candy

[Makes 36 pieces]

NOTE: These candies are also popular around St. Patrick's Day. This recipe uses actual potatoes.

Materials

1/2 cup mashed potatoes (cooked and mashed with no added ingredients)	mixing bowl
	mixing spoon
	wax paper
1 teaspoon vanilla	tray
1 pound confectioners' sugar	small plastic bag
2 tablespoons cinnamon	airtight container

Procedure

1. Combine mashed potatoes and vanilla in mixing bowl.
2. Gradually add confectioners' sugar, stirring constantly. The dough should become stiff but not sticky.
3. Remove about 1 teaspoonful of dough at a time. Form "potatoes" and set on wax paper lined tray to dry slightly.
4. Pour cinnamon into plastic bag. Place several potatoes in bag and toss gently.
5. Remove and shake off excess cinnamon.
6. Store in an airtight container.

■ Very Easy Fudge
[Makes 24 pieces]

NOTE: Fudge actually uses some of the science of crystallization. This particular recipe requires little cooking.

Materials

2 packages semisweet chocolate
(8 squares each)
1 14-ounce can sweetened condensed milk
2 teaspoons vanilla
nonstick cooking spray

microwave-safe mixing bowl
microwave oven
square pan, 8-inch by 8-inch
spoon
refrigerator

Procedure

1. Combine semisweet chocolate and sweetened condensed milk in microwave-safe mixing bowl.
2. Microwave mixture for 1 minute.
3. Stir and microwave for about 1 or 2 minutes.
4. Stir until chocolate is melted.
5. Spray square pan with nonstick cooking spray.
6. Pour fudge into pan and refrigerate until firm.
7. Cut into pieces.

■ Caramel Apples
[Makes 4 apples]

NOTE: Watch out for the caramel; make sure it is not too hot.

Materials

4 medium apples
4 wooden sticks
1 14-ounce bag caramels, unwrapped
nonstick cooking spray
2 tablespoons water

large microwave-safe mixing bowl
microwave oven
spoon
wax paper

Procedure

1. Combine unwrapped caramels and water in microwave-safe mixing bowl.
2. Microwave on high for several minutes, until caramels are melted.
3. Place a wooden stick into the stem end of each apple.
4. Dip each apple into the caramel mixture and coat.
5. Spray wax paper with nonstick cooking spray.
6. Place each apple on wax paper and cool.
7. Do not refrigerate.

■ Chocolate-Covered Pretzel Rods
[Makes 10]

NOTE: Children could dunk other foods in the melted chocolate. Dried fruits, fresh strawberries, and chunks of bananas are good candidates.

Materials

1 cup semisweet chocolate chips
10 8-inch pretzel rods
colored sprinkles in a small bowl
microwave-safe bowl

microwave oven
mixing spoon
wax paper

Procedure

1. Pour chocolate chips into the microwave-safe bowl and microwave until melted (about 1–2 minutes).
2. Stir to make mixture smooth.
3. Dunk pretzel rods into chocolate and then into the bowl of sprinkles.
4. Place on wax paper until chocolate hardens.

■ Frosting–Pudding Candy
[Makes about 30 pieces]

NOTE: Pistachio pudding would make a good flavor of this candy.

Materials

1 can prepared vanilla frosting   mixing spoon
1 small box pudding mix   wax paper
coconut   tray
2 mixing bowls   refrigerator

Procedure

1. Combine frosting and pudding in 1 mixing bowl.
2. Place coconut in the other mixing bowl.
3. Form frosting–pudding mixture into balls and roll in coconut.
4. Line tray with wax paper.
5. Place candy on tray and refrigerate until candy is firm (about 2 hours).

■ Mints
[Makes about 30 pieces]

NOTE: These mints could change with the holidays. Red food coloring could be used for Valentine's Day. Pastel colors could be used at Easter. Orange mints could be made for Halloween. Red and green colors could make Christmas mints.

Materials

3 tablespoons butter   microwave-safe bowl
3 tablespoons milk   microwave oven
2 cups confectioners' sugar   mixing spoon
food coloring   wax paper
1/2 teaspoon peppermint extract   airtight container

Procedure

1. Combine milk and butter in microwave-safe bowl.
2. Microwave mixture for about a minute or until butter melts.
3. Add confectioners' sugar and stir until combined.
4. Add food coloring and extract.
5. Drop small amounts onto wax paper to cool.
6. Store in an airtight container.

■ Berry Water Ice
[Makes 1 1/2 quarts]

NOTE: Children will say, "Nutritious and delicious."

Materials

3 cups berries (e.g., strawberries, blueberries, cranberries)
1 1/2 cups sugar
3 cups water
pot

blender
spoon
freezer-safe container
stove or heating element
freezer

Procedure

1. Pour fruit into pot and add sugar and water. Simmer until berries are soft.
2. Puree mixture in blender.
3. Pour into freezer-safe container and place in freezer.
4. Stir several times while mixture is freezing.

■ Lemon Water Ice
[Makes 1 1/2 quarts]

NOTE: Water ice is very popular. It is easy to make and quite refreshing.

Materials

1 quart water
1 1/2 cups sugar
1 cup lemon juice
pot

spoon
freezer-safe container
stove or heating element
freezer

Procedure

1. Combine water and sugar in pot. Boil for 3 minutes and cool.
2. Add lemon juice.
3. Pour into freezer-safe container and place in freezer.
4. Stir several times while mixture is freezing.

■ Frozen Fruit Treats
[Makes 6 treats]

NOTE: These treats are very nutritious. They also help children understand how water can take different forms.

Materials

1 small package strawberry- or raspberry-flavored gelatin
1/2 cup boiling water
1/2 tablespoon lemon juice
1 20-ounce can pineapple chunks with juice
2 bananas

mixing bowl
mixing spoon
blender
knife
paper cups or small plastic containers
craft sticks
freezer

Procedure

1. Mix gelatin and boiling water in mixing bowl.
2. Pour into blender. Add lemon juice and pineapple.
3. Cut bananas into small pieces and add.
4. Blend until pureed. Pour into paper cups and place in freezer.
5. When treats are almost solid, insert craft sticks. Return to freezer until solid.

■ Lemonade for a Crowd
[Makes 20 servings]

NOTE: This cool treat could follow field day activities.

Materials

4 quarts water
3 cups lemon juice
4 cups sugar
large pot or container

large mixing spoon
ladle
paper or plastic cups
ice

Procedure

1. Combine water and lemon juice in large container.
2. Stir in sugar.
3. Place ice in cups and then add lemonade.

■ Hot Cocoa for a Crowd
[Makes 16–20 servings]

NOTE: Water could be prepared ahead of time in a crock-pot; the hot cocoa could follow a cold, windy, winter recess.

Materials

2 cups powdered milk
1/3 cup cocoa
1 cup confectioners' sugar
1/3 cup powdered nondairy creamer
1/2 teaspoon salt

large mixing bowl
mixing spoon
hot cups
boiling water
ladle

Procedure

1. Combine powdered milk, cocoa, confectioners' sugar, nondairy creamer, and salt in the bowl.

2. Each cup of cocoa needs about 4 tablespoons mix and boiling water.

30
Fruit Preserves

Fruit preserves were first made to take advantage of summer's bountiful fruit. These fruit preserves were stored away until they could be used in winter. Jams, jellies, fruit butters, and marmalades are the most common types of fruit preserves. Most homemade fruit preserves are canned, but that takes time. Also, canned fruit preserves are not always safe because processing requires water baths with high temperatures. Furthermore, microbes can still enter the preserves. These freezer/refrigerator recipes require no processing, and they taste good.

The high sugar content actually keeps microbes away. Fruit should be at room temperature when any preserve is made. Pectin can be purchased in the canning section of the grocery store.

■ Raspberry–Blueberry Refrigerator Jam
[Makes 1 cup]

NOTE: Raspberries and blueberries are expensive, but they provide a great treat. Consider buying these fruits in the summertime when prices are more reasonable.

Materials

3/4 cup blueberries
3/4 cup raspberries
2 1/2 teaspoons pectin
1/2 cup sugar

2-quart microwave-safe bowl
microwave oven
mixing spoon
storage container and lid

Procedure

1. Combine blueberries, raspberries, and pectin in microwave-safe bowl.

2. Microwave for about 2 minutes until mixture is bubbling.

3. Add sugar and microwave for 3 more minutes.

4. Skim off foam and pour into the storage container.

5. Allow to cool. Add lid and refrigerate.

6. Use within 2 weeks.

■ Strawberry Freezer Jam
[Makes 4 3/4 cups]

NOTE: Make sure the strawberries are firm and unbruised.

Materials

1 quart strawberries, washed and hulled
4 cups sugar
3/4 cup water
1/3 cup pectin
mixing bowl

potato masher
pot
stove or heating element
mixing spoon
clean freezer containers with lids

Procedure

1. In mixing bowl lightly crush strawberries with potato masher and add sugar. Let this mixture sit for 10 minutes.

2. In the pot combine water and pectin. Bring to a boil and cook, stirring constantly, for 1 minute.

3. Remove pot from heat. Stir pectin–water mixture into fruit mixture and stir for 3 minutes.

4. Pour mixture into freezer containers and quickly cover with lids. Keep at room temperature for a day and then freeze.

5. After removing from freezer, keep in refrigerator. Use within 2 weeks.

■ Grape Freezer Jelly
[Makes 6 1/3 cups]

NOTE: You could make your own juice, but the bottled variety works almost as well.

Materials

3 cups unsweetened grape juice
5 1/4 cups sugar
3/4 cup water
1/3 cup pectin
mixing bowl

pot
stove or heating element
mixing spoon
clean freezer containers with lids

Procedure

1. In mixing bowl combine juice and sugar. Let this mixture sit for 10 minutes.

2. In the pot combine water and pectin. Bring to a boil and cook, stirring constantly, for 1 minute.

3. Stir pectin–water mixture into juice mixture and stir for 3 minutes.

4. Pour mixture into freezer containers and quickly cover with lids. Keep at room temperature for a day and then freeze.

5. After removing from freezer, keep in refrigerator. Use within 2 weeks.

■ Peach Freezer Jam

[Makes 5 2/3 cups]

NOTE: The lemon juice and ascorbic acid preservative keep the peaches from turning brown.

Materials

2 1/4 cups peaches, peeled, pitted, and chopped (about 2 pounds)
2 tablespoons lemon juice
1 teaspoon ascorbic acid preservative
5 cups sugar
3/4 cup water

1/3 cup pectin
mixing bowl
pot
stove or heating element
mixing spoon
clean freezer containers with lids

Procedure

1. In the mixing bowl combine peaches, lemon juice, ascorbic acid preservative, and sugar.

2. In the pot combine water and pectin. Bring to a boil and cook, stirring constantly, for 1 minute.

3. Remove the pot from the heat. Stir pectin–water mixture into fruit mixture and stir for 3 minutes.

4. Pour mixture into freezer containers and quickly cover with lids. Keep at room temperature for a day and then freeze.

5. After removing from freezer, keep in refrigerator. Use within 2 weeks.

■ Rhubarb–Cherry Freezer Jam
[Makes 2 pints]

NOTE: When we eat rhubarb we are eating the edible stem of the plant. Rhubarb leaves contain poisonous substances. Rhubarb by itself is quite tart. This recipe requires 2 days.

Materials

6 cups peeled, sliced rhubarb
4 cups sugar
1 can (21 ounces) cherry pie filling
1 package (6 ounces) cherry-flavored gelatin
mixing bowl and cover

mixing spoon
pot
stove or heating element
shallow container
clean freezer containers with lids

Procedure

1. Combine rhubarb and sugar in mixing bowl. Cover and refrigerate overnight.

2. Pour rhubarb–sugar mixture into pot and cook over medium heat until rhubarb is tender, about 10 minutes.

3. Remove from heat and add cherry pie filling and gelatin.

4. Pour into shallow container and refrigerate.

5. When mixture is cold, place in freezer containers. Cover with lids and place in freezer.

6. After removing from freezer, keep in refrigerator. Use within 2 weeks.

■ Easy Apple Butter
[Makes approximately 1 cup]

NOTE: Apple butter contains no butter. Apple butter is the outcome when chopped apples are slowly cooked and reduced to a very thick sauce. Fruits other than apples can become a fruit butter.

Materials

1 large jar applesauce
cinnamon or ground cloves or both
pot

stove or heating element
spoon

Procedure

1. Pour applesauce into pot and heat over low heat. Stir often.
2. Add spices to taste.
3. Soon applesauce should become dark and thick.
4. Cool and spread over bread.
5. Refrigerate remaining apple butter.

■ Slow-Cooker Apple Butter
[Makes 8 cups]

NOTE: Apple butter was a colonial favorite. It takes quite a while to make, but it is worth the effort. Save the apple peelings and cores (minus the apple seeds) to make apple jelly (see recipe below).

Materials

5 1/2 pounds apples, peeled, cored,
    and sliced thin
3/4 cup light brown sugar
3/4 cup white sugar
3 teaspoons cinnamon

1 teaspoon nutmeg
slow cooker
mixing spoon
potato masher or blender
clean containers with lids

Procedure

1. Combine apples, white sugar, brown sugar, and spices in slow cooker.
2. Cook on high for 1 hour.
3. Cook on low for about 10 hours.
4. Mash apples or use blender to develop uniform consistency.
5. Cool and place in containers.
6. Refrigerate and use within several weeks.

■ Apple Jelly from Apple Peelings and Cores
[Makes about 7 cups]

NOTE: This recipe uses apple peelings and cores, items usually discarded. Therefore, the jelly is reasonably priced to make, and children can learn to use everything nature provides. The finished products make great gifts.

Materials

peelings and cores without the seeds from about 20 apples (apple flesh could be used to make apple butter, apple pie, apple tarts, etc.)

water

cheesecloth and strainer

1 box pectin

8 cups sugar

large pot

mixing spoon

stove or heating element

clean jars with lids

Procedure

1. Boil apple peelings and cores (without seeds) in 6 cups water for 30 minutes.

2. Place cheesecloth in strainer. Pour contents of pot into cheesecloth. Squeeze cheesecloth and cooked apple pieces to get every bit of goodness.

3. Return apple juice to pot. Add water if necessary to have about 7 cups fluids.

4. Add pectin and bring to a boil.

5. Add sugar and boil 1 more minute.

6. Pour into containers and let cool.

7. Refrigerate and consume within several weeks.

■ Pumpkin Butter
[Makes 2 cups]

NOTE: Similar to apple butter, this fruit spread is delicious on waffles and muffins.

Materials

1 can (15 ounces) pumpkin puree
1 cup applesauce
1/3 cup packed brown sugar
1 teaspoon cinnamon
1 teaspoon ground ginger

2 1/2 tablespoons lemon juice
pot
stove or heating element
mixing spoon
clean container with lid

Procedure

1. Combine all ingredients except lemon juice in the pot.

2. Heat until mixture begins to boil, and then reduce the temperature to a simmer. Cook, stirring often, for about an hour or until pumpkin butter is very thick.

3. Add lemon juice and stir well.

4. Allow mixture to cool and then pour into the clean container. Add lid and refrigerate.

5. Use within 2 weeks.

■ Orange–Pineapple Marmalade
[Makes 4 cups]

NOTE: A marmalade uses the fruit peel as well as the fruit. This recipe uses a microwave and not a stove. Make sure that the microwave-safe container is deep, because the sauce expands as it cooks in the microwave.

Materials

2 large oranges

1 can (15 1/2 ounces) crushed pineapple with juice removed

4 cups sugar

2 1/2 tablespoons lemon juice

knife

microwave-safe container

microwave oven

mixing spoon

clean freezer containers with lids

Procedure

1. Wash and cut oranges. Remove seeds and membranes separating the segments.

2. Cut oranges into small pieces.

3. Combine oranges with other ingredients in microwave-safe container and microwave at high, stirring occasionally, for about 8 minutes until mixture begins to boil.

4. Microwave at high for another 2–3 minutes.

5. Pour marmalade into the clean freezer containers with lids.

6. Cool for 4 hours and then freeze.

7. The marmalade can remain in the freezer for a year, or it can stay in the refrigerator for about 2 weeks.

31
Syrups

Real maple syrup is made from the sap of maple trees. In early spring the sap is tapped and collected. About 25 gallons of sap are boiled down to make 1 gallon of maple syrup. Making real maple syrup in a classroom is not possible, but fruit syrups are easy to make. These syrups go well with waffles, pancakes, and French toast. They could also act as toppings for ice cream and yogurt desserts.

■ Blueberry Syrup
[Makes 1 cup]

NOTE: Health experts believe blueberries are extremely nutritious. The berries are low in calories, and they contain high amounts of antioxidants.

Materials

2 cups fresh or frozen blueberries
1/2 cup water
1/2 cup sugar
2 teaspoons lemon juice

pot
stove or heating element
mixing spoon

Procedure

1. In the pot combine 1 cup blueberries, water, sugar, and lemon juice.
2. Cook over medium heat, stirring constantly, until sugar dissolves.
3. Bring to a boil and then reduce heat to simmer. Cook, stirring constantly, for about 20 minutes until mixture is thick.
4. Add remaining blueberries and simmer for about 3 more minutes.
5. Remove pot from stove and allow syrup to cool.
6. Syrup can be stored in the refrigerator for about a week.

■ Grape Syrup
[Makes 1 3/4 cups]

NOTE: This recipe is probably the easiest and cheapest to make.

Materials

1 cup grape jam or jelly
1/2 cup light corn syrup
1/4 cup water

pot
stove or heating element
mixing spoon

Procedure

1. In the pot combine grape jam or jelly, corn syrup, and water. Stir over medium heat until jelly melts.
2. Serve warm.
3. Syrup can be stored in the refrigerator for about a week.

■ Cranberry Syrup
[Makes 2 cups]

NOTE: Cranberries by themselves are quite tart.

Materials

2 1/2 cups cranberry juice
3/4 cup light corn syrup
1/2 cup sugar

pot
stove or heating element
mixing spoon

Procedure

1. In the pot combine cranberry juice, light corn syrup, and sugar. Stir over medium heat until sugar dissolves.
2. Bring to a boil and let simmer for about a half hour.
3. Serve warm.
4. Syrup can be stored in the refrigerator for about a week.

■ Peanut Butter Syrup
[Makes 1 1/2 cups]

NOTE: Peanut butter syrup should not be eaten by anyone who is allergic to nuts.

Materials

4 tablespoons butter or margarine
1/3 cup creamy peanut butter
1 cup maple-flavored syrup
1/2 cup water

pot
stove or heating element
mixing spoon

Procedure

1. Melt butter in the pot over low heat.
2. Stir in peanut butter.
3. Add syrup and water and combine.
4. Bring to a boil, stirring constantly.
5. Simmer for about 5 minutes or until slightly thickened.
6. Serve warm.
7. Syrup can be stored in the refrigerator for about a week.

■ Brown Sugar Syrup
[Makes 1 2/3 cups]

NOTE: This syrup, similar to a caramel sauce, is particularly good over ice cream.

Materials

2 cups packed brown sugar   pot
1 cup water   stove or heating element
1 1/2 teaspoons vanilla   mixing spoon

Procedure

1. Combine brown sugar and water in the pot. Stir over medium heat until brown sugar melts.
2. Bring to a boil and reduce heat. Cook, stirring constantly, on low heat for about 5 minutes.
3. Remove from heat and add vanilla.
4. Serve warm.
5. Syrup can be stored in the refrigerator for about a week.

■ Apple–Maple Syrup
[Makes 1 cup]

NOTE: Children could find out how apple cider differs from apple juice.

Materials

1 cup apple cider   stove or heating element
1 cup maple-flavored syrup   mixing spoon
pot

Procedure

1. Combine cider and syrup in the pot.
2. Bring to a boil and reduce heat. Cook, stirring constantly, on low heat for about 15 minutes until syrup is thick.
3. Serve warm.
4. Syrup can be stored in the refrigerator for about a week.

■ Orange Syrup
[Makes 1 cup]

NOTE: Children could find out how maple syrup is different from maple-flavored syrup. Several grades of real maple syrup can be found. How do the grades differ from one another?

Materials

1 cup maple-flavored syrup
1 orange, peeled and sliced
pot

stove or heating element
mixing spoon

Procedure

1. Combine syrup and orange slices in the pot.
2. Bring to a boil and reduce heat. Cook, stirring constantly, on low heat for about 5 minutes.
3. Remove from heat and remove orange slices.
4. Serve warm.
5. Syrup can be stored in the refrigerator for about a week.

32
Pickles, Sauerkraut, Mustards, and Horseradish

Pickling, a method of preserving food, dates back to very early cultures. The high salt and vinegar content discourages the growth of microbes. Today the ability to refrigerate and freeze fruits and vegetables has made pickling less necessary. True pickling requires canning, which takes a great deal of time and preparation. Canning can also be dangerous, because high temperatures are necessary to kill bacteria. The following recipes do not require canning, but they do give children a sense of how to preserve foods.

■ Refrigerator Pickles
[Makes 6 cups]

NOTE: The vinegar and salt form brine, a way to preserve vegetables. The solution becomes so acidic and so salty that bacteria cannot grow in it.

Materials

6 cups thinly sliced cucumbers
1 cup sliced onions
1 1/2 cups sugar
1/2 teaspoon salt
1/2 teaspoon mustard seed
1/2 teaspoon celery seed

1/2 teaspoon turmeric
1 1/2 cups white vinegar
large, very clean jar with lid
pot
stove or heating element
mixing spoon

Procedure

1. Layer cucumbers and onions in the large jar.

2. Combine the other ingredients in the pot.

3. Bring mixture to a boil, stirring until sugar is dissolved.

4. Remove mixture from heat and pour over cucumbers and onions.

5. Cool for about 2 hours and then cover tightly and refrigerate for several days.

6. Eat the pickles within 2 weeks.

■ Refrigerator Dill Pickles
[Makes about 9 cups]

NOTE: Dill is an herb that children could easily grow indoors and then transplant outdoors in the springtime.

Materials

10 cucumbers
1 cup white vinegar
1 cup water
1 tablespoon salt
2 tablespoons dill
1 clove garlic, crushed
1/2 teaspoon mustard seeds

knife
pot
stove or heating element
mixing spoon
large crock or glass bowl with lid or
    plastic wrap

Procedure

1. Wash cucumbers and slice with knife.
2. Combine water, vinegar, and salt in pot and bring to a boil. This will form a brine.
3. Place a layer of dill, crushed garlic, and mustard seeds on the bottom of the crock.
4. Place cucumbers on top of the herbs.
5. Add another layer of dill.
6. Cover cucumbers with brine mixture.
7. Cover and place in the refrigerator for several days.
8. Eat the pickles within 2 weeks.

■ Refrigerator Bread and Butter Pickles
[Makes 1 quart]

NOTE: The sugar and vinegar give these pickles both a sweet and sour taste.

Materials

1 quart cucumbers
1 sliced onion
1 cup sugar
1 teaspoon turmeric
1/2 teaspoon mustard seeds
1 cup white vinegar

1 1/4 teaspoon salt
large pot
stove or heating element
mixing spoon
large crock or glass bowl with lid or
    plastic wrap

Procedure

1. Wash and slice cucumbers.

2. In the large pot combine sugar, turmeric, mustard seed, vinegar, and salt. Bring to a boil.

3. Add cucumbers and sliced onion and bring the mixture to a boil again.

4. Remove pot from heating element and allow mixture to cool for about a half hour.

5. Pour mixture into crock and cover. Refrigerate for several days.

6. Eat the pickles within 2 weeks.

■ Sauerkraut
[Makes about 1 gallon]

NOTE: Sauerkraut is a way of preserving cabbage. The salt creates an environment that keeps bacteria away. Note that the volume of cabbage greatly reduces when it becomes sauerkraut. Pickling salt is different from table salt because pickling salt contains no additives (such as iodine). Pickling salt is also a finer grain. It can be found in the canning supplies section of the grocery store.

Materials

about 5 pounds of cabbage
knife
3 1/2 tablespoons pickling salt
large crock or jar
wooden spoon

clean cloth
plate big enough to cover mouth of
    crock or jar
a weight to keep plate down on top of
    crock or jar

Procedure

1. Shred cabbage.

2. Layer cabbage and salt in crock or jar.

3. Cover with the cloth, plate, and weight. Place crock in a dry, room-temperature location.

4. Each day for about 10 days skim the scum from the top of the mixture.

5. Thoroughly clean the cloth to prevent mold.

6. In about 10 days fermentation should be complete, and the sauerkraut can be tasted.

7. Refrigerate and consume within several days.

■ All-Purpose Mustard
[Makes 2/3 cup]

NOTE: Homemade (or school-made) mustard is likely to be more watery and stronger flavored than commercial mustards. Any mustard should mellow for at least a full day before being tasted.

Materials

2 tablespoons coarsely ground brown mustard seeds
2 tablespoons coarsely ground yellow mustard seeds
1/4 cup mustard powder
1/4 cup cold water

2 tablespoons white vinegar
1/2 teaspoon salt
mixing bowl
mixing spoon
jar with lid

Procedure

1. Combine mustard seeds, mustard powder, and water together in mixing bowl. Let stand for 10 minutes.

2. Stir in vinegar and salt.

3. Pour into jar and cap. Refrigerate for at least 24 hours before serving.

4. Consume the mustard within 2 weeks.

■ Sweet-Hot Mustard
[Makes about 1 1/2 cups]

NOTE: Different types of mustard exist. Some are stronger than others. Children could grind mustard seeds and try out different types.

Materials

4 ounces dry mustard powder   double boiler
3 large eggs   water for double boiler
1 cup apple cider vinegar   stove or heating element
1 cup brown sugar   mixing spoon
blender   jar and lid

Procedure

1. Combine all ingredients in blender and blend until smooth.
2. Pour mixture into the top of a double boiler. Place water in the bottom of the double boiler.
3. Bring almost to a boil and then reduce heat. Cook until thickened.
4. Pour mixture into jar and let cool. Screw on lid.
5. Let mellow for at least 24 hours.
6. Consume mustard within 2 weeks.

■ Honey Mustard
[Makes about 1 1/4 cups]

NOTE: The honey gives this mustard a bit of sweetness.

Materials

1/3 cup dry mustard   1 teaspoon Worcestershire sauce
1/2 cup honey   mixing bowl
1/2 cup brown sugar   mixing spoon
1/4 cup vinegar   clean jar with lid
1/4 cup vegetable oil

Procedure

1. Combine all ingredients in mixing bowl.
2. Pour into jar and refrigerate for at least 24 hours.
3. Consume the mustard within about 2 weeks.

■ Homemade Horseradish
[Makes about 1 cup]

NOTE: Horseradish has been consumed for thousands of years. Horseradish is one of the bitter herbs served at Passover meals. The Greeks used ground horseradish as a salve for back pain. American colonists grew it in their gardens and used it to flavor stews and many types of meats.

Materials

1 large, fresh horseradish root (no soft spots or green coloring)
vegetable peeler
knife
grater
wax paper
vinegar or sour cream (optional)
clean jar with lid

Procedure

1. Peel off outer layer of the root.
2. Cut horseradish root into slices and remove the fibrous core.
3. Grate the horseradish root over a piece of wax paper.
4. Serve as is or mix with vinegar or sour cream.
5. Store in a jar with the lid screwed on tightly.
6. Refrigerate. Consume the horseradish within 2 weeks.

33
Miscellaneous

As the name indicates, these recipes and formulas do not fall into any category. However, the following are still important, useful, or just plain fun!

■ Spray "Shellac"
[Makes 2 cups]

NOTE: This "shellac" is inexpensive, easy to make, and can be used by children. It provides a nice, shiny finish. Spraying is best done outside.

Materials

1 1/2 cups water
1/2 cup white glue

Spray bottle

Procedure

1. Pour water into spray bottle.

2. Add white glue and replace spray bottle nozzle.

3. Shake spray bottle.

4. To use, take art piece to be shellacked outside. Spray with "shellac" and let dry.

■ Applesauce–Cinnamon Decorations

NOTE: Use holiday cookie cutters to make decorations to hang on a tree. Use heart-shaped cookie cutters to make Valentine's Day gifts. Make sure the recipients of these gifts know the decorations are not edible. Cinnamon can be expensive, so plan ahead and look for sales.

Materials

1 cup applesauce	cookie cutters
1 4-ounce container ground cinnamon	straw
mixing bowl	drying rack
mixing spoon	spatula
wax paper	ribbon
rolling pin	

Procedure

1. Mix applesauce and cinnamon in mixing bowl until a stiff dough forms.
2. Turn mixture onto wax paper and roll to 1/4-inch thickness.
3. Cut into shapes or use cookie cutters. Using the straw, make a hole at the top of the decoration for the ribbon.
4. Lay decorations on drying rack.
5. Flip decorations over twice a day for 10 days so that edges will not curl.
6. When decorations are thoroughly dry, thread a ribbon through the hole on each one and knot it.

■ Pumpkin Pie Spice Decorations

NOTE: These items make great Thanksgiving decorations and table favors. They are not edible.

Materials

3/4 cup ground cinnamon   extra cinnamon to dust surface
1/4 cup ground nutmeg   cookie cutter
1 teaspoon ground ginger   straw
5 tablespoons white glue   drying rack
mixing bowl   spatula
mixing spoon   ribbon
wax paper   refrigerator
rolling pin

Procedure

1. Combine cinnamon, nutmeg, ginger, and glue in mixing bowl.

2. Add enough water to make a stiff dough.

3. Refrigerate for 3 hours.

4. Cover work area with wax paper. Sprinkle extra cinnamon on wax paper. Place dough on wax paper.

5. Knead dough until pliable. Using rolling pin, roll dough to a 1/2-inch thickness.

6. Use cookie cutter to cut out dough. With the straw, make a hole at the top of each decoration.

7. Lay decorations on drying rack to dry.

8. Using the spatula, flip decorations over twice a day for 1 week. This step allows the decorations to dry evenly. They may curl if they are not turned.

9. When decorations are dry, thread holes with ribbon and hang.

■ Spatter Painting

NOTE: Smocks are a good idea for this project. Children like to watch the randomness of the spatters.

Materials

newspaper
toothbrush
piece of window screening
paints

blank paper
4 small blocks or boxes (1 1/2–2
    inches tall) to form props for the
    screen

Procedure

1. Cover work area with newspaper.
2. Place blank paper on newspaper.
3. Place a block or box at each corner of the paper.
4. Place window screening above paper and on the blocks.
5. Load toothbrush with paint.
6. Brush toothbrush across screening. Paint will spatter over the paper.
7. To make a negative print, cut out a shape such as a heart. Place the paper heart on the blank paper. Spatter the paper and carefully remove the shape. The spatter paint will outline the shape.

■ Glue Sun Catchers

NOTE: These decorations take a day or two to dry. The tricky part is removing the dried glue from the frame.

Materials

wax paper
cookie cutters
straws cut into 2-inch segments
paper cups
white glue

food coloring
toothpicks
glitter, beads, or other small decorations
ribbon

Procedure

1. Cover work area with wax paper.

2. Place cookie cutters on wax paper.

3. In paper cups, combine white glue with food coloring. Use toothpicks to blend colors.

4. Place a straw segment upright near the top of each cookie cutter. The straw will provide a hole from which to hang the decoration.

5. Pour some colored glue mixture into the center of each cookie cutter. Make sure the straw is not disturbed.

6. Add glitter and other decorations if desired.

7. Allow glue to dry. Remove cookie cutters and wax paper.

8. Remove or cut away the straw. String ribbon through the hole and hang.

■ Sugar–Glitter Sun Catchers
[Makes about 8]

NOTE: These can be made to match the seasons.

Materials

1 cup sugar	cookie tray
4 teaspoons glitter	wax paper
2–3 teaspoons water	cookie cutters
mixing bowl	glue
mixing spoon	8 12-inch pieces of thin ribbon

Procedure

1. Combine sugar and glitter in the mixing bowl.
2. Add enough water to make the mixture stick together.
3. Line cookie sheet with wax paper.
4. Place a cookie cutter on wax paper.
5. Fill cookie cutter with about 2 teaspoons of the material. Make sure to spread material to a uniform thickness.
6. Gently remove cookie cutter.
7. Repeat with other cookie cutters and the rest of the material.
8. Allow shapes to dry overnight.
9. Glue the thin ribbon onto the backs so that the sun catchers can be displayed.

■ Tie-Dying

NOTE: Oh, no! The '60s are back! Actually, tie-dyeing was practiced over a thousand years ago in Asia and Africa. Although children can still tie-dye T-shirts, they can also fashion book covers, lamp shades, scarves, and many other items. The natural dyes featured in Chapter 8 can be used, but they may not be colorfast. Also, different fabrics dye at different rates. Cotton seems the most dependable. Silk dyes the quickest. Synthetic fabrics are not always predictable. Everyone should wear rubber gloves.

Materials

fabric	rubber gloves
newspaper	sink with running water
household dye	bucket of soapy water
bucket to hold dye	rubber bands

Procedure

1. Wash the fabric first to remove any sizing.
2. Cover area with newspaper.
3. Prepare dye according to manufacturer's directions. Make sure dye is hot.
4. Dunk fabric in soapy water. The soap will help retain the dye.
5. Bunch the fabric at locations and tie with rubber bands. Use at least 2 rubber bands at each tie, and make sure the bands are very tight.
6. Put on rubber gloves. Place fabric in dye. The longer it remains, the darker the final color will be.
7. When the desired shade is reached, remove fabric from dye and rinse under running water.
8. Remove rubber bands to examine pattern and hang fabric to dry.
9. If more than 1 color is desired, band and dye the fabric in 1 color. Let dry. Then band and dye in another color. Always work from lighter colors to darker colors.
10. Objects such as marbles can be knotted into the fabric with rubber bands. This will create a consistently repeating pattern.

■ Batik

NOTE: Batik making may be an older technique than tie-dyeing. It probably originated in eastern Asia. Everyone should wear rubber gloves. A hot iron removes the wax at the end of the project.

Materials

newspaper
aluminum foil
fabric, preferably cotton
large, flat pan of cool water
paraffin
coffee can
electric frying pan

paintbrush
household dye
large, flat pan to hold dye
sink with running water
towels
iron
ironing board

Procedure

1. Cover work area with newspaper.
2. Place a large piece of aluminum foil on top of the newspaper. The foil will keep the paraffin-coated fabric from sticking to the newspaper.
3. Place fabric on top of foil.
4. Place paraffin in coffee can. Place coffee can in electric frying pan. Add water to the electric frying pan and turn on the heat. The coffee can and electric frying pan act as a double boiler to prevent the wax from catching fire.
5. When wax has melted, dip brush in the wax. "Paint" a picture on the fabric with the wax. Dip the brush into wax often to paint picture. Now the fabric is a batik.
6. Place batik in cool water for a few minutes to harden wax.
7. Prepare dye according to manufacturer's directions. Place dye in the large, flat pan.
8. Put on rubber gloves. Remove batik from cool water and place it in the dye. Remember that the longer the batik sits in the dye, the darker the color will be.
9. When the desired shade is reached, remove batik from dye. Rinse in running water.
10. Place batik on towel and remove excess water.
11. Heat the iron. Place batik on ironing board. Cover fabric with newspaper. Apply hot iron.
12. The paraffin will melt into the newspaper and away from batik. Change newspaper and repeat process until wax is fully removed.
13. A batik crackle can be made by applying wax to the entire fabric surface. Place batik into very cold water to chill the wax. Crumple fabric to crack the wax. Then place batik in dye. The dye will enter the cracks. Remove wax as per steps 9 through 12.

■ String Decorations

NOTE: Only thin string works well with this project. Starch cannot support the weight of heavy string. Crochet thread, especially the variegated kind, produces interesting results. Make small string decorations for Christmas tree ornaments or Easter eggs. Make large string ornaments to liven up the walls for parties. For another project, cover only the lower portion of the balloon. When the balloon is popped and removed, the string will form a basket.

Materials

liquid starch
small balloon
medium-size bowl
crochet thread or other light string
(15 years for a balloon inflated to

the size of a light bulb; 30 yards
for a balloon inflated to the size of
a cantaloupe)

Procedure

1. Blow up balloon to desired size and secure with a knot.
2. Pour liquid starch into bowl. Soak string in liquid starch for at least 30 minutes.
3. Remove string from starch and wrap it haphazardly around the balloon. Make sure all major portions of the balloon are covered.
4. Hang the balloon from under a door way or the underside of a table. Allow liquid starch to dry.
5. Pop and remove balloon. The remaining string decoration is light and intriguing.

■ Paper Jewelry

NOTE: Children can make necklaces from these beads in no time. This project incorporates crafts into a study of ancient Egypt.

Materials

colorful pages from catalogs and magazines

scissors

glue

yarn or elastic

manufactured beads (optional)

Procedure

1. Cut colorful pages into isosceles triangles. The base of each triangle can range from 1/2 inch to 1 1/2 inches in length. The longer the base of each triangle, the longer the resulting bead will be. The 2 equal sides of each triangle should be between 2 and 3 inches. The longer the sides, the thicker the bead will be.

2. Starting at the base edge, roll triangle to the pointed end. Leave a space in the center of each roll.

3. Glue tip of triangle to rest of rolled bead. Let dry.

4. Following steps 1 through 3, make beads of different sizes and colors.

5. String beads on elastic or yarn to make necklaces or bracelets. Add other types of beads to the pattern if desired. Tie ends of the elastic or yarn to complete the project.

■ Rubber Stamp Ink

[Makes 1]

NOTE: Children could have several different colors of stamp pad inks available. These inks are much cheaper than commercial forms.

Materials

2 packages powdered sugar-free drink mix

2 tablespoons glycerin

3/4 teaspoon water

disposable bowl

disposable spoon

stamp pad

Procedure

1. Combine powdered drink mix, glycerin, and water in disposable bowl.

2. Pour over stamp pad.

■ Rubber Stamp Pad
[Makes 1]

NOTE: Make sure children tightly cover pad container to keep ink from drying out.

Materials

1 sponge

1 plastic container such as a cottage cheese container with lid

scissors

water

stamp pad ink

Procedure

1. Cut sponge to fit bottom of container.

2. Dampen sponge with a bit of water. Do not soak sponge.

3. Add stamp pad ink.

4. Tightly cover pad at end of project.

■ Potato Stamps

NOTE: Children can use potato stamps to make repeating patterns. Use stamps on large pieces of paper to make wrapping paper.

Materials

potato

knife

pencil

paints and paintbrushes

paper

water

paper towels

Procedure

1. Wash and dry potato.

2. Cut potato in half.

3. Draw a simple shape (such as a heart) on the cut surface of one potato half.

4. Using the knife, cut away the potato from around the design. This leaves a design raised on the cut surface.

5. Paint the raised area with paints.

6. Immediately stamp the potato on the paper.

7. Depending on the effect desired, the potato may need to be repainted every time. However, it could be used several times before needing to be repainted.

■ Vegetable and Fruit Stamps

NOTE: An unusual still life can be created using these natural stamps. Mushrooms have particular appeal.

Materials

firm vegetables or fruits, such as mushrooms, cauliflower, apples, or pears

knife

paints and paintbrushes

paper

water

paper towels

Procedure

1. Wash and dry vegetables and fruits.
2. Cut vegetables and fruits in half.
3. Paint the cut, flat areas of the fruits and vegetables with paints.
4. Immediately stamp the vegetables and fruits on the paper.
5. Depending on the effect desired, the vegetables and fruits may need to be repainted every time. However, they could be used several times before needing to be repainted.

■ Pine Needle Paintbrushes

[Makes 10]

NOTE: Children can find out how commercial paintbrushes are made.

Materials

green pine needles

10 stocky twigs or small branches

10 pieces of string or yarn, each 1 foot long

paint

paper

Procedure

1. Gather a number of pine needles and place around the end of the twig.
2. Tie in place with yarn or string.
3. Experiment with paint.

■ Newspaper Paintbrushes
[Makes 10]

NOTE: Children can decide which paintbrushes work better, newspaper or pine needle.

Materials

10 sturdy plastic straws   paint
about 2 sheets newspaper   paper
10 pieces of string or yarn, each 1
    foot long

Procedure

1. Tear newspaper into about 12 small strips.

2. Place around the end of the straw.

3. Tie in place with yarn or string.

4. Experiment with paint.

■ Colored Sand
[Makes 1 cup]

NOTE: This sand is far cheaper than colored sand purchased at craft stores.

Materials

1 cup sand or salt   airtight container
3 teaspoons powdered tempera paint

Procedure

1. Pour sand and powdered tempera paint into airtight container.

2. Cover and shake.

■ Decorative Plate
[Makes 1 plate]

NOTE: Dollar stores are good sources for the glass plates.

Materials

clear glass plate
sheet white paper
scissors
markers or crayons
pencil

sponge brush
1/4 cup white glue
1/4 cup water
disposable cup

Procedure

1. Turn plate upside down and place white paper on the bottom.
2. Trace the bottom shape onto the white paper and cut out the shape.
3. Use the markers or crayons to create a decorative scene.
4. Pour equal amounts of water and white glue into the disposable cup and combine.
5. Use the sponge brush to coat the bottom of the plate.
6. Place the picture onto the bottom of the plate so that when the plate is turned over the scene appears.
7. Let dry.
8. Coat with another layer or two of the white glue–water mixture.
9. Let dry again.

■ Quick Beanbags
[Makes 1]

NOTE: Children can decide which filler is best.

Materials

1 sock
1 cup filler such as dried beans or
dried barley

1 piece of string about 15 inches long

Procedure

1. Pour filler into sock.
2. Tie string around sock opening.
3. Toss gently and have fun.

■ Saltwater Chalk
[Makes enough for 1 child]

NOTE: Throw away the chalk after the activity. Different colored chalks make a nice finished product.

Materials

disposable cup

2 teaspoons salt

1/4 cup warm water

1 stick chalk

pale construction paper

mixing spoon

Procedure

1. Combine salt and warm water in disposable cup.

2. Dip chalk into warm salt water and let it remain there for about a minute.

3. Take it out and start to draw.

■ Crepe Paper Raffia
[Makes 1]

NOTE: This project takes a delicate hand.

Materials

1 crepe paper streamer

spool

Procedure

1. Twist the end of the crepe paper.

2. Carefully insert the end of crepe paper streamer through the hole of the spool.

3. Turn and pull length of crepe paper through the spool.

■ Sand Art
[Makes 1]

NOTE: The more crayon the better the final piece.

Materials

1 piece sandpaper
crayons
oven

cookie sheet lined with aluminum foil
hot pad

Procedure

1. With the crayons create a scene on the piece of sandpaper. Press hard with the crayons.
2. Place sandpaper on aluminum foil-lined cookie sheet.
3. Bake in the oven at 350°F for about 20 minutes.
4. Remove from oven and let cool.

■ Tissue Paper "Stained Glass Windows"
[Makes 1]

NOTE: These really brighten up a classroom. Liquid starch can be purchased at grocery stores.

Materials

colored tissue paper
liquid starch
wax paper

scissors
paintbrush
small container filled with water

Procedure

1. Cut small pieces of tissue paper and arrange on a piece of wax paper to form a picture.
2. Lift each piece one at a time and paint underside with liquid starch.
3. Place tissue paper back on wax paper.
4. Clean brush each time when changing tissue paper color.
5. Let dry and hang from window.

■ Dyed Eggs Using Tissue Paper
[Makes 6]

NOTE: The eggs should not be eaten because some of the tissue paper dye may not be safe to consume.

Materials

6 hard boiled eggs
small pieces of brightly colored tissue paper

old paint brush
water in a small container
2 paper towels

Procedure

1. Paint eggs with water.
2. Place small pieces of tissue paper all over the eggs.
3. Paint tissue paper with more water so that tissue paper is very wet.
4. Place eggs on paper towels and allow them to dry.
5. The tissue paper will fall off, and the eggs will be colored.

■ Indoor "Mud Puddles"
[Makes 1 large batch that can be used by several children]

NOTE: Save the coffee grounds from the faculty room coffee pot for several days prior to this activity.

Materials

12 cups used and dry coffee grounds
1 container dry oatmeal

1 container salt
1 large, flat pan

Procedure

1. Pour dry coffee grounds into the large, flat pan.
2. Allow children to play in the coffee grounds "mud puddles."
3. Allow children to add oatmeal or salt to coffee grounds and explore the differences.

■ Transfer Liquid

[Makes a bit more than 1/4 cup, enough for quite a long time]

> NOTE: Use this mixture to transfer pictures from magazines or newspapers to projects or other types of paper. Comics are a great source of transfer material. This project is best done outside.

Materials

2 tablespoons soap powder (not detergent) or shavings from a bar of soap

1/4 cup hot water

1 tablespoon turpentine*

disposable container, such as cottage cheese container

disposable mixing spoon

jar with screw top

picture

paper

old paintbrush

smooth rock or other object about the size of an egg

*Turpentine should not be consumed. It is also highly flammable.

Procedure

1. Combine soap powder and hot water in disposable container.

2. Add turpentine.

3. Allow to cool and pour into jar. Screw on the top.

4. When ready to use, shake well. Paint a layer of liquid to the picture to be transferred. Wait about 10 seconds.

5. Place the paper over the picture and rub the area with the smooth rock. Wait a few seconds and then remove the paper. The picture should now be on the paper.

6. Sometimes pictures can be used again.

■ Mobile
[Makes 1]

NOTE: Balancing the objects takes a bit of patience.

Materials

coat hanger
2 pencils or small dowels
string or yarn
8 objects to hang from mobile

scissors
hole punch
glue

Procedure

1. Punch a hole in the top of each of the eight objects.

2. Tie string or yarn through the holes. Leave a generous amount of string attached to each object.

3. Tie 2 strings to the ends of 1 pencil or dowel. Firmly attach by adding a bit of glue.

4. Attach 2 strings to the ends of the other pencil or dowel. Again firmly attach with a bit of glue.

5. Tie a piece of string or yarn to the middle of each pencil and secure with glue.

6. Tie all the strings to the bottom of the coat hanger. Use various lengths of string or yarn so that the objects hang at various lengths.

7. Hold the coat hanger by the hook. Make sure the eight objects balance on the coat hanger. Move some to another area if necessary.

8. Glue the strings on to the coat hanger to keep everything in place.

Bibliography

Bawden, Juliet. *The Art and Craft of Papier Mâché*. San Francisco: Chronicle Books, 1995.

Brown, Jordan. *Crazy Concoctions: A Mad Scientist's Guide to Messy Mixtures*. Watertown, MA: Charlesbridge Publishers, 2011.

Churchill, E. Richard, Louis V. Loeschnig, and Muriel Mandell. *365 Simple Science Experiments with Everyday Materials*. New York: Black Dog & Leventhal Publishers, Inc., 2013.

Cobb, Vicki. *See for Yourself: More than 100 Amazing Experiments for Science Fairs*. New York: Skyhorse Publishing, 2010.

Earnest, Don, ed. *Homemade: How to Make Hundreds of Everyday Products Fast, Fresh, and More Naturally*. Pleasantville, NY: The Reader's Digest Association, Inc., 2008.

Hannah, Sue. *Crafty Concoctions: 101 Shiny, Sparkly, Spooky, Wrinkly, Puffy, Funny, Easy (and Money Saving) Craft Supplies You Can Make*. New York: Meadowbrook Press, 2003.

Hauser, Jill Frankel. *Super Science Concoctions: 50 Mysterious Mixtures for Fabulous Fun*. Nashville: Williamson Books, 2007.

Reader's Digest. *For the Birds: Easy-to-Make Recipes for your Feathered Friends*. Pleasantville, NY: The Reader's Digest Association, Inc., 2010.

Rhatigan, Joe, and Veronika Alice Gunter. *Cool Chemistry Concoctions: 50 Formulas that Fizz, Foam, Splatter & Ooze*. New York: Lark Books, 2005.

Roth, Sally. *The Backyard Bird Feeder's Bible: The A-to-Z Guide to Feeders: Seed Mixes, Projects, and Treats*. Emmaus, PA: Rodale Book Readers' Service, 2000.

Satler, Helen Roney. *Recipes for Art and Craft Materials*. New York: Lothrop, Lee & Shepard; William Morrow, 1973.

Thomas, John E., and Danita Thomas. *The Ultimate Book of Kid Concoctions*. Strongsville, OH: The Kid Concoctions Company, 2000.

TIME for Kids Big Book of Science Experiments: A Step-by-step Guide. New York: Time for Kids Books, 2011.

About the Author

DIANA F. MARKS consults with teachers, parents, children, and school districts to help enhance learning and develop curricula. A retired teacher of gifted students, she enjoys creating innovative and interesting activities for children. Her published works include Libraries Unlimited's *Let's Celebrate Today: Calendars, Events, and Holidays, Volumes 1 and 2* and *Children's Book Award Handbook*. She has a master's degree in education.